THE SOCIAL CONSTRUCTION OF WHAT?

IAN HACKING

THE SOCIAL CONSTRUCTION OF WHAT?

HARVARD UNIVERSITY PRESS
CAMBRIDGE, MASSACHUSETTS
AND LONDON, ENGLAND • 1999

Printed in the United States of America

Library of Congress Cataloging-in-Publication Data

Hacking, Ian.
The social construction of what? / Ian Hacking.
p. cm.
Includes bibliographical references and index.
ISBN 0-674-81200-X (alk. paper)
1. Knowledge, Sociology of. I. Title.
BD175.H29 1999
121—dc21 98-46140
CIP

CONTENTS

For Catherine

PREFACE

Social construction is one of very many ideas that are bitterly fought over in the American culture wars. Combatants may find my observations rather like the United Nations resolutions that have little effect. But a lot of other people are curious about the fray going on in the distance. They are glad to hear from a foreign correspondent, not about the wars, but about an idea that has been cropping up all over the place.

I have seldom found it helpful to use the phrase "social construction" in my own work. When I have mentioned it I have done so in order to distance myself from it. It seemed to be both obscure and overused. Social construction has in many contexts been a truly liberating idea, but that which on first hearing has liberated some has made all too many others smug, comfortable, and trendy in ways that have become merely orthodox. The phrase has become code. If you use it favorably, you deem yourself rather radical. If you trash the phrase, you declare that you are rational, reasonable, and respectable.

I used to believe that the best way to contribute to the debates was to remain silent. To talk about them would entrench the use of the phrase "social construction." My attitude was irresponsible. Philosophers of my stripe should analyze, not exclude. Even in the narrow domains called the history and the philosophy of the sciences, observers see a painful schism. Many historians and many philosophers won't talk to each other, or else they talk past each other, because one side is so contentiously "constructionist" while the other is so dismissive of the idea. In larger arenas, public scientists shout at sociologists, who shout back. You almost forget that there are issues to discuss. I have tried to get some perspective on established topics in the field. More interesting are some openings to new ideas that have not yet been examined.

Labels such as "the culture wars," "the science wars," or "the Freud wars" are now widely used to refer to some of the disagreements that plague contemporary intellectual life. I will continue to employ those labels, from time to time, in this book, for my themes touch, in myriad ways, on those confrontations. But I would like to register a gentle protest. Metaphors influence the mind in many unnoticed ways. The willingness to describe fierce disagreement in terms of the metaphors of war makes the very existence of real wars seem more natural, more inevitable, more a part of the human condition. It also betrays us into an insensibility toward the very idea of war, so that we are less prone to be aware of how totally disgusting real wars really are.

And now for acknowledgments. Usually I work for years on something, pretty much by myself, aided by interested students at my own university. These chapters, first presented as lectures or seminars, are, for me, unusual, because the ideas have been worked out in public, above all with students at the University of Toronto. My first thoughts about social construction were written down for Irving Velody, who asked me for a piece to go in the book of an English conference that I did not attend. A much revised version now serves as Chapter 2. Then I was asked to talk about social construction in its former heartlands, the New School of Social Research in New York, and Frankfurt University, where the nonlecture Chapter 2 became a real lecture. I ended up doing lectures all over the place: as Henrietta Harvey lecturer at Memorial University, Newfoundland (Chapter 1); the George Myro lecture, Berkeley, California (Chapter 3); two lectures (Chapters 3 and 4) at the Institut de l'Histoire des Sciences et Philosophie et Technique, Paris I (Sorbonne). Chapter 4 is an extended version of the John Coffin Memorial Lecture, in London, and Chapter 3 was given as a follow-up seminar. In Tokyo, Chapter 1 served for a seminar at the Ecole des Hautes Etudes en Sciences Humaines, Tokyo, and Chapter 3 for research workers at Fuji Xerox, Tokyo, and also at Kyoto University.

Chapters 1, 2, and 4 formed a final set of lectures at Green College in the University of British Columbia. The idea of three talks came at the beginning of these travels, when Richard Ericson, the President of Green College, in a single conversation, both suggested I give a set of lectures at the college a couple of years later, and said that my book on multiple personality, *Rewriting the Soul,* was a classic of social constructionism.

I was as taken aback by the second remark as I was honored by the first, so it is fitting that the final version of this evolution was delivered a couple of years later, at Green College, in January 1998. I wish particularly to thank Ernie Hamm for ensuring that everything went smoothly there.

Chapters 1–4 are, then, extended versions of four lectures on fairly different aspects of social construction.

Chapter 2 is substantially revised from "On Being More Literal about Construction," in *The Politics of Constructionism,* ed. I. Velody and R. Williams (London: Sage, 1998), reprinted by permission of Sage Publications Ltd. Parts of Chapter 4 appeared as "Taking Bad Arguments Seriously," *London Review of Books,* 21 August 1997. Chapter 5 is shortened and adapted from "World-making by Kind-making: Child Abuse for Example," in *How Classification Works: Nelson Goodman among the Social Sciences,* ed. Mary Douglas and David Hull (Edinburgh: Edinburgh University Press, 1992). Chapter 6 appeared in essentially its present form as "Weapons Research and the Form of Scientific Knowledge," *Canadian Journal of Philosophy* (1997), Supplementary Vol. 12: 327–348. Chapter 8, revised here, first appeared as "Was Captain Cook a God?", *London Review of Books,* 7 September 1995. I thank the various publishers for permission to use the texts.

Chapter 7 has been adapted from a lecture for high school science teachers in Portugal, organized by Fernando Gil, under the auspices of the Ministry of Education. It is more old-fashioned than the other chapters because it explains some traditional philosophy of science, though it also introduces contemporary science studies. It is old-fashioned in another way too. Dr Johnson refuted Bishop Berkeley's immaterialist philosophy by kicking a rock, and today one reads that Maxwell's Equations are as real as—rocks. I could not resist taking that seriously. Why not think about geology and social construction? The example is built around a very common kind of rock, dolomite. Happily the example, based on current research done in Zurich by Dr Judith McKenzie and her collaborators, manages to touch on many a topic, including early forms of life, and maybe, if you want to speculate a little, life on Mars.

My ideas have not so much changed during the travels that produced chapters 1–4 and 7, as been clarified. Every single talk exposed many things that I had not thought about. Ignorance and confusion remain, but the time has come to stop wandering. Collectively my audiences were participants in the making of this book. Some contributions from individuals are flagged in the notes, but to all a hearty thanks. Some people say that the culture wars have temporarily destroyed the possibility of friendly discussion and scholarly collaboration. What do I think about that? I have always wanted to use in print a word I learned from long-ago comic strips, so now I can. *Pshaw!*

WHY ASK WHAT?

What a lot of things are said to be socially constructed! Here are some construction titles from a library catalog:

Authorship (Woodmansee and Jaszi 1994)
Brotherhood (Clawson 1989)
The child viewer of television (Luke 1990)
Danger (McCormick 1995)
Emotions (Harré 1986)
Facts (Latour and Woolgar 1979)
Gender (Dewar, 1986; Lorber and Farrell 1991)
Homosexual culture (Kinsman 1983)
Illness (Lorber 1997)
Knowledge (MacKenzie 1981, Myers 1990, Barrett 1992, Torkington 1996)
Literacy (Cook-Gumperz 1986)
The medicalized immigrant (Wilkins 1993)
Nature (Eder 1996)
Oral history (Tonkin 1992)
Postmodernism (McHale 1992)
Quarks (Pickering 1986)
Reality (Berger and Luckmann 1966)
Serial homicide (Jenkins 1994)
Technological systems (Bijker, Hughes, and Pinch 1987)
Urban schooling (Miron 1996)
Vital statistics (Emery 1993)
Women refugees (Moussa 1992)
Youth homelessness (Huston and Liddiard 1994)
Zulu nationalism (Golan 1994)

Not to mention Deafness, Mind, Panic, the eighties and Extraordinary science (Hartley and Gregory 1991, Coulter 1979, Capps and Ochs 1995, Grünzweig and Maeirhofer 1992, Collins 1982). Individual people also qualify: at a workshop on teenage pregnancy, the overworked director of a Roman Catholic welfare agency said: "And *I* myself am, of course, a social construct; each of us is."[1] Then there is experience: "Scholars and activists within feminism and disability rights have demonstrated that the experiences of being female or of having a disability are socially constructed" (Asche and Fine 1988, 5f).

My alphabetical list is taken from titles of the form *The Social Construction of X,* or *Constructing X.* I left *X* out of my alphabet for lack of a book, and because it allows me to use *X* as a filler, a generic label for what is constructed. Talk of social construction has become common coin, valuable for political activists and familiar to anyone who comes across current debates about race, gender, culture, or science. Why?

For one thing, the idea of social construction has been wonderfully liberating. It reminds us, say, that motherhood and its meanings are not fixed and inevitable, the consequence of child-bearing and rearing. They are the product of historical events, social forces, and ideology.[2] Mothers who accept current canons of emotion and behavior may learn that the ways they are supposed to feel and act are not ordained by human nature or the biology of reproduction. They need not feel quite as guilty as they are supposed to, if they do not obey either the old rules of family or whatever is the official psycho-pediatric rule of the day, such as, "you must bond with your infant, or you both will perish."[3]

Unfortunately social construction analyses do not always liberate. Take anorexia, the disorder of adolescent girls and young women who seem to value being thin above all else. They simply will not eat. Although anorexia has been known in the past, and even the name is a couple of hundred years old, it surfaced in the modern world in the early 1960s. The young women who are seriously affected resist treatment. Any number of fashionable and often horrible cures have been tried, and none works reliably. In any intuitive understanding of "social construction," anorexia must in part be some sort of social construction. It is at any rate a transient mental illness (Hacking 1998a), flourishing only in some places at some times. But that does not help the girls and young women who are suffering. Social construction theses are liberating chiefly for those who are on the way to being liberated—mothers whose consciousness has already been raised, for example.

For all their power to liberate, those very words, "social construction," can work like cancerous cells. Once seeded, they replicate out of hand. Consider Alan Sokal's hoax. Sokal, a physicist at New York University, published a learned pastiche of current "theory" in *Social Text,* an important academic journal for literary and cultural studies (Sokal 1996a). The editors included it in a special issue dedicated to the "science wars." In an almost simultaneous issue of *Lingua Franca,* a serious variant of *People* magazine, aimed at professors and their ilk, Sokal owned up to the mischief (Sokal 1996b). Sokal's confession used the term "social construction" just twice in a five-page essay. Stanley Fish (1996), dean of "theory," retorted on the op-ed page of the *New York Times.* There he used the term, or its cognates, sixteen times in a few paragraphs. If a cancer cell did that to a human body, death would be immediate. Excessive use of a vogue word is tiresome, or worse.

In a talk given in Frankfurt a few days after the story broke in May of 1996, I said that Sokal's hoax had now had its fifteen minutes of fame. How wrong I was! There are several thousand "Sokal" entries on the Internet. Sokal crystallized something very important for American intellectual life. I say American deliberately. Many of Sokal's targets were French writers; and Sokal's own book on these topics was first published in French (Bricmont and Sokal 1997a). That in turn produced two French books, both with the French word *impostures* in their titles (Jenneret 1998, Jurdant 1998). The European reaction has, however, remained bemused rather than concerned. Plenty of reporting, yes, but not much passion. In late 1997 Sokal had little prominence in Japan, although the most informative Sokal website anywhere had just opened in Japanese cyberspace.[4] Students of contemporary American mores have an obligation to explain the extraordinary brouhaha that Sokal provoked in his own country. My aim is not to give a social history of our times explaining all that, but to analyze the idea of social construction, which has been on the warpath for over three decades before Sokal. Hence I shall have almost nothing to say about the affair. Readers who want a polemical anthology of American writing siding with Sokal may enjoy Koertge (1998).

RELATIVISM

For many people, Sokal epitomized what are now called the "science wars." Wars! The science wars can be focused on social construction.

One person argues that scientific results, even in fundamental physics, are social constructs. An opponent, angered, protests that the results are usually discoveries about our world that hold independently of society. People also talk of the culture wars, which often hinge on issues of race, gender, colonialism, or a shared canon of history and literature that children should master—and so on. These conflicts are serious. They invite heartfelt emotions. Nevertheless I doubt that the terms "culture wars," "science wars" (and now, "Freud wars") would have caught on if they did not suggest gladiatorial sport. It is the bemused spectators who talk about the "wars."

There is, alas, a great deal of anger out there that no amount of lightheartedness will dispel. Many more things are at work in these wars than I can possibly touch on. One of them is a great fear of relativism. What is this wicked troll? Clear statements about it are hard to find. Commonly, people suspected of relativism insist they are not haunted by it. A few, such as the Edinburgh sociologists of science, Barry Barnes and David Bloor (1982), gladly accept the epithet "relativist." Paul Feyerabend (1987), of "anything goes" fame, managed to describe some thirteen versions of relativism, but this attempt at divide-and-rule convinced no one.

I think that we should be less highbrow than these authors. Let us get down to gut reactions. What are we afraid of? Plenty. There is the notion that any opinion is as good as any other; if so, won't relativism license anything at all? Feminists have recently cautioned us about the dangers of this kind of relativism, for it seems to leave no ground for criticizing oppressive ideas (Code 1995). The matter may seem especially pressing for third-world feminists (Nanda 1997).

Then there is historical revisionism. The next stage in the notorious series of holocaust denials might be a book entitled *The Social Construction of the Holocaust*, a work urging that the Nazi extermination camps are exaggerated and the gas chambers fictions. No one wants a relativism that tells us that such a book will, so far as concerns truth, be on a par with all others. My own view is that we do not need to discuss such issues under the heading of relativism. The question of historical revisionism is a question of how to write history.[5] Barnes and Bloor (1983, 27) make plain that relativist sociologists of their stripe are obliged to sort out their beliefs and actions, using a critical version of the standards of their own culture. Feyerabend's last words (1994) were that every culture is one culture, and we ought to take a stand against oppression anywhere. And I ended my own contribution to a book on rationality and relativism by quoting Sartre's last words explaining why the Jewish and Islamic traditions played no part in his thought: they did not for the simple reason that they were no part of his life (Hacking 1983).

There are more global bogeymen. Intellectuals and nationalists are frightened of religious fundamentalism in India, Israel, the Islamic world, and the United States. Does not relativism entail that any kind of religious fundamentalism is as good as any kind of science?

Or maybe the real issue is the decline of the West (in the United States, read America). Decline is positively encouraged by some social constructionists, is it not? Sometimes people focus on the loss of tradition and resent "multiculturalism." That is one fear that I cannot take seriously, perhaps because the word was in use, in a purely positive way, in Canada long before it got taken up in the American culture wars. My goodness, where I live my provincial government has had a Minister of Multiculturalism for years and years; I'm supposed to be worried about that?

Relativism and decline are real worries, but I am not going to address them directly. It is good to stay away from them, for I cannot expect successfully to dispel or solve problems where so many wise heads have written so many wise words without effect. More generally, I avoid speculating further on the profound malaise that fuels today's culture wars. I am at most an unhappy witness to it, saddened by what it does.

DON'T FIRST DEFINE, ASK FOR THE POINT

Social construction talk has recently been all the rage. I cannot hope to do justice to all parties. I shall take most of my examples from authors who put social construction up front, in their titles. They may not be the clearest, most sensible, or most profound contributors, but at any rate they are self-declared. So what are social constructions and what is social constructionism? With so many inflamed passions going the rounds, you might think that we first want a definition to clear the air. On the contrary, we first need to confront the point of social construction analyses. Don't ask for the meaning, ask what's the point.

This is not an unusual situation. There are many words or phrases of which the same thing must be said. Take "exploitation." In a recent book about it, Alan Wertheimer (1996) does a splendid job of seeking out necessary and sufficient conditions for the truth of statements of the form

"*A* exploits *B*." He does not quite succeed, because the point of saying that middle-class couples exploit surrogate mothers, or that colleges exploit their basketball stars on scholarships—Wertheimer's prized examples—is to raise consciousness. The point is less to describe the relation between colleges and stars than to change how we see those relations. This relies not on necessary and sufficient conditions for claims about exploitation, but on fruitful analogies and new perspectives.

In the same way, a primary use of "social construction" has been for raising consciousness.[6] This is done in two distinct ways, one overarching, the other more localized. First, it is urged that a great deal (or all) of our lived experience, and of the world we inhabit, is to be conceived of as socially constructed. Then there are local claims, about the social construction of a specific *X*. The *X* may be authorship or Zulu nationalism. A local claim may be suggested by an overarching attitude, but the point of a local claim is to raise consciousness about something in particular. Local claims are in principle independent of each other. You might be a social constructionist about brotherhood and fraternity, but maintain that youth homelessness is real enough. Most of this book is about local claims. That is why I began with the question, "The social construction of what?" and opened with a list of whats. The items in my alphabetical list are so various! Danger is a different sort of thing from reality, or women refugees. What unites many of the claims is an underlying aim to raise consciousness.

AGAINST INEVITABILITY

Social construction work is critical of the status quo. Social constructionists about *X* tend to hold that:

(1) *X* need not have existed, or need not be at all as it is. *X*, or *X* as it is at present, is not determined by the nature of things; it is not inevitable.

Very often they go further, and urge that:

(2) *X* is quite bad as it is.
(3) We would be much better off if *X* were done away with, or at least radically transformed.

A thesis of type (1) is the starting point: the existence or character of *X* is not determined by the nature of things. *X* is not inevitable. *X* was brought into existence or shaped by social events, forces, history, all of which could well have been different. Many social construction theses at once advance to (2) and (3), but they need not do so. One may realize that something, which seems inevitable in the present state of things, was not inevitable, and yet is not thereby a bad thing. But most people who use the social construction idea enthusiastically want to criticize, change, or destroy some *X* that they dislike in the established order of things.

GENDER

Not all constructionists about *X* go as far as thesis (3) or even (2). There are many grades of commitment. Later on I distinguish six of them. You can get some idea of the gradations by thinking about feminist uses of construction ideas. Undoubtedly the most influential social construction doctrines have had to do with gender.[7] That was to be expected. The canonical text, Simone de Beauvoir's *The Second Sex,* had as its most famous line, *On ne naît pas femme: on le devient;* "One is not born, but rather becomes, a woman" (de Beauvoir 1949, II, 1; 1953, 267). It also suggested to many readers that gender is constructed.[8]

Previous toilers in the women's movements knew that power relations needed reform, but many differences between the sexes had a feeling of inevitability about them. Then feminists mobilized the word "gender." Let *X* = gender in (1)–(3) above. Feminists convinced us (1) that gendered attributes and relations are highly contingent. They also urged (2) that they are terrible, and (3) that women in particular, and human beings in general, would be much better off if present gender attributes and relations were abolished or radically transformed. Very well, but this basic sequence (1)–(3) is too simplistic. There are many differences of theory among feminists who use or allude to the idea of construction.[9]

One core idea of early gender theorists was that biological differences between the sexes do not determine gender, gender attributes, or gender relations. Before feminists began their work, this was far from obvious. Gender was, in the first analyses, thought of as an add-on to physiology, the contingent product of the social world. Gender, in this conception, is "a constitutive social construction: . . . Gender should be understood as a social category whose definition makes reference to a broad network of social relations, and it is not simply a matter of anatomical differences" (Haslanger 1995, 130).[10]

Many constructionist uses of gender go beyond this add-on approach. Naomi Scheman (1993, ch. 18) inclines to functionalism about gender. That is, she thinks that the category of gender is in use among us to serve ends of which members of a social group may not be aware, ends which benefit some and only some members of the group. The task is to unmask these ends, to unmask the ideology. When Scheman says that gender is socially constructed, she means in part that it motivates visions in which women are held to be essentially, of their very nature, subject to male domination.

Scheman wants to reform the category of gender. Judith Butler is more rebellious. She insists that individuals become gendered by what they do—a favored word is "performance." She rejects the notion that gender is a constructed add-on to sexual identity. Male and female bodies are not givens. My body is, for me, part of my life, and how I live that life is part of the determination of what kind of body I have. "Perhaps this construct called 'sex' is as culturally constructed as gender . . . with the consequence that the distinction between sex and gender turns out to be no distinction at all" (Butler 1990, 7).

We may here be reminded, but only for a moment, of Thomas Laqueur's (1990) observations of how differently the sex organs have been represented in, among other things, Western medical texts of the past millennium. Butler is not discussing such systems of knowledge about the body. They have, of course, limned some possibilities for perception of self, and influenced possibilities for acting, living. But her concern goes far beyond Laqueur's. The systems of knowledge that he presents all assume that sex is physiological, a given prior to human thought. They differ about what is given. Butler questions how we get the idea of that given. Older notions of gender do not help answer such questions. "How, then," she asks, "does gender need to be reformulated to encompass the power relations that produce the effect of a prediscursive sex and so conceal that very operation of discursive production?" Thus she wants at least to revise early feminist notions of gender, and as I read her, wants to mature away from talk of construction and proceed to a more complex analysis that would, perhaps, shed the word "construction" altogether.

Butler cites as an ally an author whose work is revolutionary. Monique

Wittig (1992, 9) repudiates the feminist tradition that affirms the power of being woman. The entire set of sexual and gender categories should be overthrown. According to Wittig, the lesbian is an agent of revolution because she lives out a refusal to be either man or woman.

Scheman, to use a ranking I shall elaborate later, is a *reformist* constructionist who wants to *unmask* some ideology. Butler's published work is what I call *rebellious,* while Wittig's is *revolutionary.* But do not imagine that all feminists are hospitable to social construction talk. I suggested that Butler distances herself from it, preferring concepts of greater precision and subtlety. Jeffner Allen seems to have avoided it from the start. She thinks that too much of such talk gets caught up in banal and narcissistic postmodern fascinations with mere texts. It diverts attention away from the basics, like wage inequalities. Quite in opposition to Wittig, she suggests that it might be a good idea to refashion a specifically feminine sensitivity. She can be caustic about the idea that she, herself, is socially constructed. Which society did you have in mind? she asks (Allen 1989, 7).

WOMEN REFUGEES

What is said to be constructed, if someone speaks of the social construction of gender? Individuals as gendered, the category of gender, bodies, souls, concepts, coding, subjectivity, the list runs on. I have used gender as an example to get us started. It is far too intense a topic to fit any easy schematism. So let me venture a small clarification using a less controversial item from my alphabetical list of titles—women refugees.

Why would someone use the title *The Social Construction of Women Refugees* (Moussa 1992), when it is obvious that women are refugees in consequence of a sequence of social events? We all think that the world would be a better place if there were no women refugees. We do not mean that the world would be better if women were simply unable to flee intolerable conditions, or were killed while so doing. We mean that a more decent world would be one in which women were not driven out of their homes by force, threats of force, or at any rate did not feel so desperate they felt forced to flee. When X = Women refugees, propositions (1), (2), and (3) are painfully obvious. What, then, could possibly be the point of talking about the social construction of women refugees?

To answer, we must, as always, examine the context. The discussion does not spring from an ideal: let no women be forced to flee. The perspective of Moussa (1992) is that of the host country (in this case Canada, which in recent years, for all its faults, has had the refugee policy that most closely approximates that of United Nations resolutions on refugees). What is socially constructed is not, in the first instance, the individual people, the women refugees. It is the classification, *woman refugee.* Moussa addresses the idea of "the woman refugee" as if that were a kind of human being, a species like "the whale." She argues that this way of classifying people is the product of social events, of legislation, of social workers, of immigrant groups, of activists, of lawyers, and of the activities of the women involved. This kind of person, as a specific *kind* of person, is socially constructed. Or simply: the *idea* of the woman refugee is constructed.

IDEAS IN THEIR MATRICES

"Idea" is shorthand, and a very unsatisfactory shorthand it is too. The trouble is that we want some general way to make the distinction needed, not just for X = women refugees, but for a host of other items said to be socially constructed. "Idea" may have to serve, although more specific words like "concept" and "kind" are waiting in the wings. I do not mean anything curiously mental by "idea." Ideas (as we ordinarily use the word) are usually out there in public. They can be proposed, criticized, entertained, rejected.

Ideas do not exist in a vacuum. They inhabit a social setting. Let us call that the *matrix* within which an idea, a concept or kind, is formed. "Matrix" is no more perfect for my purpose than the word "idea." It derives from the word for "womb," but it has acquired a lot of other senses—in advanced algebra, for example. The matrix in which the idea of the woman refugee is formed is a complex of institutions, advocates, newspaper articles, lawyers, court decisions, immigration proceedings. Not to mention the material infrastructure, barriers, passports, uniforms, counters at airports, detention centers, courthouses, holiday camps for refugee children. You may want to call these social because their meanings are what matter to us, but they are material, and in their sheer materiality make substantial differences to people. Conversely, ideas about women refugees make a difference to the material environment (women refugees are not violent, so there is no need for guns, but there is a great need for paper, paper, paper). Materiel influences the people (many of whom have no comprehension of that paper, paper, paper, the different offices, the uniforms). Sheer matter, even the color of the paint on the walls, can gradually replace optimistic hope by a feeling of impersonal grinding oppression.

This discussion of ideas and classification takes for granted the obvious, namely that they work only in a matrix. But I do want to emphasize what in shorthand I call the *idea* of the woman refugee, that classification, that kind of person. When we read of the social construction of *X*, it is very commonly the idea of *X* (in its matrix) that is meant. And ideas, thus understood, do matter. It can really matter to someone to be classified as a woman refugee; if she is not thus classified, she may be deported, or go into hiding, or marry to gain citizenship. The matrix can affect an individual woman. She needs to become a women refugee in order to stay in Canada; she learns what characteristics to establish, knows how to live her life. By living that life, she evolves, becomes a certain kind of person (a woman refugee). And so it may make sense to say that the very individuals and their experiences are constructed within the matrix surrounding the classification "women refugees."

Notice how important it is to answer the question "The social construction of what?" For in this example *X* does not refer directly to individual women refugees. No, the *X* refers first of all to the woman refugee as a kind of person, the classification itself, and the matrix within which the classification works. In consequence of being so classified, individual women and their experiences of themselves are changed by being so classified.

This sounds very complicated. But the logical point is simple. Women in flight are the product of social conditions in their homelands. It would be stupid to talk about social construction in that context, because social circumstances so manifestly provoke the fear of staying home and the hope of succor in another land. But since, in Canada, *woman refugee* may seem a straightforward and rather inevitable way of classifying some people, there is indeed a point to claiming that the classification is far from inevitable. One can also argue that this contingent classification, and the matrix within which it is embedded, changes how some women refugees feel about themselves, their experiences, and their actions. Hence in that indirect way people themselves are affected by the classification—and, if you like, the individual herself is socially constructed as a certain kind of person.

A PRECONDITION

Notice how thesis (1)—*X* need not have existed—sets the stage for social construction talk about *X*. If everybody knows that *X* is the contingent upshot of social arrangements, there is no point in saying that it is socially constructed. Women in flight, or at the immigration barrier, are there as a result of social events. Everyone knows that, and only a fool (or someone who likes to jump on bandwagons) would bother to say that they are socially constructed. People begin to argue that *X* is socially constructed precisely when they find that:

(0) In the present state of affairs, *X* is taken for granted; *X* appears to be inevitable.

In my example, the concept of the woman refugee seems inevitable, once you have the practices of nationality, immigration, citizenship, and women in flight who have arrived in your country begging asylum. The author of a book on the social construction of women refugees is saying no, the concept, and the matrix of rules, practices, and material infrastructure in which it is embedded, are not inevitable at all.

Statement (0) is not an assumption or presupposition about *X*. It states a precondition for a social constructionist thesis about *X*. Without (0) there is no inclination (aside from bandwagon jumping) to talk about the social construction of *X*. You can confirm this by scrolling down the *A* through *Z* above. You do not find books on the social construction of banks, the fiscal system, cheques, money, dollar bills, bills of lading, contracts, tort, the Federal Reserve, or the British monarchy. These are all contractual or institutional objects, and no one doubts that contracts and institutions are the result of historical events and social processes. Hence no one urges that they are socially constructed. They are part of what John Searle (1995) calls social reality. His book is titled *The Construction of Social Reality,* and as I explained elsewhere (Hacking 1997), that is not a *social* construction book at all.

I left out *J* in my alphabetical list. I could have gone from "constructing" to "inventing," with *Inventing Japan: The Making of a Post-War Civilization* (Chapman 1991). The title is possibly a pun, in the manner of the book called *Inventing Leonardo* (Turner 1993): postwar Japan is inventive and invented. (There are two books titled *Inventing Women,* Panabaker 1991, and Kirkup and Keller 1992; one is about women inventors, and one is about how roles for women in science were invented.)

The book about Japan is a history book with a thesis. It argues that modern Japan is a wholly new phenomenon. The common claim that Japan is deeply rooted in ancient tradition is, says the author, false. Regardless of the truth of his thesis, the phenomena he presents are obviously social phenomena, but no one files this book with the social construction literature. This is partly because, if the topic is contemporary Japan, the nation, then condition (0) is not satisfied. No one could think that the modern nation arose inevitably.

On the other hand, if the topic is the *idea* of Japan, that does seem more inevitable. Take some books with similar titles; *Inventing America* (Rabasa 1993); *Inventing Australia* (White 1981); *Inventing Canada* (Zeller 1987); *Inventing Europe* (Delanty 1995); *Inventing New England* (Brown 1995); *Inventing India* (Crane 1992); *Inventing Ireland* (Kiberd 1996). The 1991 *Inventing Japan* appears, in retrospect, to have participated in an early 1990s orgy of inventions, composed for people who think that the idea of nation or region *X*, with all its connotations in fiction and stereotypes, is pretty inevitable. In short, for people who act as if condition (0) were satisfied.

Since the Federal Reserve is so obviously the upshot of contingent arrangements, a book titled *The Social Construction of the Federal Reserve* would likely be silly; we would suspect someone was trying to cash in on the cachet of "social construction." But we can imagine a startling work, *The Social Construction of the Economy.* Every day we read that the economy is up or down, and we are supposed to be moved to fear or elation. Yet this splendid icon, the economy, was hard to find on the front pages of newspapers even forty years ago. Why are we so unquestioning about this very idea, "the economy"? One could argue that the idea, as an analytic tool, as a way of thinking of industrial life, is very much a construction. It is not the economy of Sweden in the year 2000 that one argues is a social construction (obviously it is that; condition (0) is not satisfied). Instead, that seemingly inevitable and unavoidable idea, *the economy,* may be argued to be a social construct.

A more terrifying creature than the economy has emerged from the fiscal woods: the deficit. That is familiar as the great political slogan of reaction of the early 1990s. Another bestseller could well be *Constructing the Deficit.* Of course the deficit was brought into being by a great deal of borrowing in the course of recent history; that is not what would be in question. The topic of this imagined bestseller would be the construction of the *idea* of the deficit. We can foresee the argument. The idea of the deficit was constructed as a threat, a constraining element in the lives of many, an instrument for the restoration of the hegemony of capital, and for the systematic and ruthless unweaving of the social net. It was constructed as a device for encouraging poor people willingly to consign themselves to yet more abject poverty.

In what follows I shall lay great emphasis on the difficult distinction between object and idea. Starting point (0) does not hold for the objects (the deficit or the economy). Obviously our present economy and our present deficit were not inevitable. They are the contingent upshot of historical events. Starting point (0) does, in contrast, hold for the ideas of the economy or the deficit; these ideas, with many of their connotations, seem inevitable.

THE SELF

Statement (0) helps clarify one very popular site for social construction analyses: "the self." I have a little trouble here. We seldom encounter anyone talking about "the self," except for rather highbrow conversation. This is quite unlike the situation with women refugees, a down-to-earth and practical topic. Our English word "self" works better as a suffix (herself) and a prefix (self-importance) than as a substantive. That is significant, but I do not want to practice linguistic philosophy here. We have to accept a situation in which many scholars contentedly discuss the self.

The history of modern philosophy contains many discussions that can induce talk about constructing the self. All of them (to foreshadow a theme developed in the next chapter) go back to Kant, and his visions of the way in which both the moral realm and the framework for the material realm are constructed.

Take existentialism. Readers of Camus or the early Sartre can form a picture of a self with absolutely no center, a self that constructs itself by free acts of will. The constructed self must, however, accept agonizing responsibility for that which it has constructed. Later, Sartre with greater awareness of Heidegger and Karl Jaspers, thought of the self as being constructed in a social matrix. This suggests a genuine distinction in which some constructions of the self are social, and some are not. Thus May (1992, 3) writes of a view, which he calls "social existentialism," and which he finds "worth reviving"; one "which derives from Heidegger, Jaspers and the later Sartre [and which] sees the self as a social construct, as a function of the interplay of history, social conditioning, and the chosen behavior of the individual person." This is the very view, quoted earlier, expressed by the overworked director of the welfare agency: "And *I* myself am, of course, a social construct; each of us is."

The point of saying *social* construct is to contrast it with individualist, and in the case of Camus and early Sartre almost solipsist, construction of the self. Note that the quasi-solipsist construction of the self is rather naturally called construction. We have the picture of a self step by step coalescing through a sequence of free acts, each of which must build on the self built up by preceding free acts. Conversely, the "interplay of history, social conditioning, and the chosen behavior of the individual person" can hardly be called *construction* at all. Only a somewhat unreflective usage—the result of rote and repetition—of terms like "social construct" would prompt one to call the resultant self a social construct. Social product, product of society, yes, but construct?

Some people find the social construction of the self repugnant for quite the opposite reason. Far from thinking of the self as beginning in a centerless Sartrian vacuum, they identify "the self" with a religious, mystical, metaphysical, or transcendental vision of the soul. Selves have essences, and, except in superficial and accidental ways, they are not constructs. Sartre, early and late, thought this was simply a mistake, so here we have a profound philosophical disagreement masquerading under the label of construction, pro or con.

There is yet another ground of objection, more empiricist than the last. Today's English-language traditions of political theory emphasize individual liberty and individual rights. Human beings are thought of as self-subsistent atoms who enter into relationships with other human beings. Enlightenment philosophies of the social contract theories had such a background, as do present-day game-theoretic approaches to ethics. Such pictures invite us to think that first there are individual "selves," and then there are societies. That has been a fruitful model in terms of which to think about justice, duty, government, and law. People who subscribe to this vision or strategy find talk of social construction suspect.

Others, who began by thinking in that way, come to realize that, despite their upbringing and the assumptions of much of the political discourse that governs the societies they inhabit, the atomistic presocial self is a harmful myth. They then find it rather liberating to proclaim that the self is a construct. That is one reason we have heard so much about the social construction of the self. It comes from people who once found the notion of a presocial self natural, even inevitable. They feel that condition (0) has been satisfied: in the present state of affairs, the atomistic self is taken for granted; it appears to be inevitable. (And it isn't inevitable at all.)

Some thinkers find atomistic visions of human nature to be obviously false. Rather, we are born into a society, educated by it, and our "selves" are sculpted out of biological raw material by constant interaction with our fellow humans—not to mention the material environments that our extended families and larger communities have made. Charles Taylor (1995) is one distinguished philosopher who takes this stance. He uses anti-Enlightenment German authors as his authorities in this connection—what he calls the Hamann-Herder-Humboldt axis. For such a thinker, there seems very little point in talking about the social construction of the self, because condition (0) is not satisfied. The self (whatever that is imagined to be) does not seem in the least inevitable.

ESSENTIALISM, ABOUT RACE, FOR EXAMPLE

Statement (0) says that *X* is taken for granted; *X* appears to be inevitable. This formulation is deliberately weak and vague. Often social construction theses are advanced against a stronger background. They are used to undermine the idea that *X* is essential, even that *X* has an "essence." Debates about the self furnish an obvious example. For something more down to earth, take race. Obviously, essentialism is an especially strong form of background assumption (0). If a person's race is an essential element of a person's being, then race is not inevitable only in the present state of affairs. It is inevitable, period, so long as there are human beings with anything like our evolutionary history on the face of the earth. Hence the anthropologist Lawrence Hirschfeld (1996) contrasts "constructionist" and "essentialist" views about race. Essentialists (usually more implicit than explicit in their beliefs) hold that one's race is part of one's "essence."

Very often essentialism is a crutch for racism, but it need not be. Hirschfeld, deeply imbued with recent cognitive science approaches to developmental psychology, argues from his experimental data that children have an innate disposition to sort people according to races, and are programmed to take an essentialist attitude to certain classifications of people, an attitude which is strongly reinforced by cultural background. This "psychological essentialism" is proposed, in part, to explain the prevalence of concepts of race and the ease with which they can be conscripted for racism. Hirschfeld argues that unqualified constructionism about race clouds our view.

Out-and-out social constructionism about race is far more politically correct than essentialism. Most anti-racialist writing denounces essentialist attitudes to race. Anthony Appiah and Amy Gutman do so in their recent book about color (1996). They may not use the label "social construction" much, but they are regularly grouped among social constructionists about race.[11]

Essentialism comes to the fore in many other highly controversial sites. Feminists have opposed views of gender and even sex as essential properties. Some debates about the nature of homosexuality can be cast as essentialism versus constructionism. The book edited by Stein (1990b), which is widely respected, is a collection of papers half of which incline to constructionism, and half to essentialism. Stein himself (1990a) produced a succinct analysis of the issues. As elsewhere, it is important to sort out the various "whats" that may be said to be socially constructed—or essential. Homosexual individuals? Homosexual culture? Homosexual practices? Homosexual genes? The homosexual as a kind of person?

As a philosopher I am, in respect of essences, an heir of John Locke and John Stuart Mill, skeptical of the very idea of essence. I am too much of their party to discuss essentialism impartially. But we do not need to. It suffices to work under the weaker umbrella notion of inevitability used in statements (0) and (1). For our purposes, essentialism is merely the strongest version of inevitability.

Notice, however, that "essentialism" is not purely descriptive. Most people who use it use it as a slur word, intended to put down the opposition. I cannot recall anyone standing up and saying, "I am an essentialist about race." Not even (so far as I know) Philippe Rushton, who presents book upon book of scientific arguments that race is an objective category that sorts human beings into three essential classes, color-coded as black, white, and yellow. He believes that members of each class tend to have a large number of characteristics distinctive of the class of which they are members, such as levels of intelligence, sex drive, athletic prowess, sociability, and so on. (e.g. Rushton 1995). In short, races have what the philosophers call essences. Nevertheless, although Rushton stands up and says the most amazing things in public, even he does not say, "I am an essentialist about race."

EMOTIONS

Emotions provide yet another field for disagreement. Some students of the subject think that there are basic, pan-cultural emotions, expressed on human faces, recognized by human beings of every culture, and produced in brain centers, all of them determined by evolutionary history. Others argue that emotions and their expression are quite specific to a social and linguistic group. Paul Ekman (1998), one of the most dedicated universalists, has provided a personal account of the controversy before the social construction era. His opponents then were those mighty figures of a yet earlier generation, Margaret Mead and Geoffrey Bateson. Nowadays the issues have been translated into social construction talk. When people say that the emotions are socially constructed, or that the emotion of grief, say, is a social construct, they do not mean that the *idea* of the emotions, or of grief is constructed, but that the emotions themselves, grief itself, are social constructs. But the word "construct" has lost all force here. In fact the "emotion" entry in my alphabetical list refers to Rom Harré's *The Social Construction of the Emotions* (1986). He told me that the original title was to be *The Social Production of the Emotions,* but the publisher insisted on *Construction.* believing that would sell more copies of the book. His later anthology, Harré and Parrott (1996), includes many essays by divers hands about social construction. The authors argue that emotions vary from culture to culture, that the character of grief has changed in Western culture and is changing today, and that the physiological expressions of emotion vary from group to group. They argue, in various ways, that how we describe emotions affects how the emotions are experienced.

The exact expression of such a thesis depends, of course, on what the author thinks emotions are. Griffiths (1997, ch. 6) notes that "There are two very different models of the social construction of emotion in the literature." There is a *social concept* model, according to which emotions are inherently cognitive and conceptual, and are the concepts peculiar to a social group, formed by the culture of that group. Then there is a *social role* model, in which "an emotion is a transitory social role (a socially constituted syndrome)" (Averill 1980, 312, quoted by Griffiths). In these discussions, the label "social construction" is more code than description. There is no literal sense in which either the Victorian concept or the Victorian role of grief was *constructed* during Her Most Britannic Majesty's long reign. "Social construct" is code for not universal, not part of pan-cultural human nature, and don't tread on me with those heavy hegemonic (racist, patriarchal) boots of yours. Griffiths sensibly contends that the "insights of social constructionism [about the emotions] are perfectly compatible with what is known about the evolutionary [and therefore biological, pre-cultural] basis of emotion" (p. 138). Since we are not talking about anything that is literally constructed, it is not obvious that these insights are best couched in terms of construction talk at all. But there is the residual force of starting point (0). Constructionists about the emotions do start by feeling that "In the present state of affairs, the emotions are taken for granted; the emotions and our expressions of them appear to be inevitable."

GRADES OF COMMITMENT

Very roughly, the gradations of constructionist commitment arise from increasingly strong reactions to (1), (2), and (3) below: (1) was the claim that *X* is not inevitable; (2) that *X* is a bad thing; and (3) that the world would be a better place without *X*. Here are names for six grades of constructionism.

Historical
Ironic
Reformist Unmasking
Rebellious
Revolutionary

The least demanding grade of constructionism about *X* is historical. Someone presents a history of *X* and argues that *X* has been constructed in the course of social processes. Far from being inevitable, *X* is the contingent upshot of historical events. A historical constructionist could be quite noncommittal about whether *X* is good or bad. How does historical "social" constructionism differ from history? Not much, a matter of attitude, perhaps.

The next grade of commitment takes an *ironic* attitude to *X*. *X*, which we thought to be an inevitable part of the world or of our conceptual architecture, could have been quite different. We are nevertheless stuck with it, it forms part of our way of thinking which will evolve, perhaps, in its own way, but about which we can do nothing much right now. The name used for this stance takes its cue from Richard Rorty's title, *Contingency, Irony and Solidarity.* Irony about *X* is the recognition that *X* is highly contingent, the product of social history and forces, and yet something we cannot, in our present lives, avoid treating as part of the universe in which we interact with other people, the material world, and ourselves.

The ironist, we feel, is a kibitzer, a powerful intellect, well able to understand the architecture of the world that pertains to *X,* but ironically forced to leave it much as it is. A third grade of commitment takes (2) seriously: *X* is quite bad as it is. Agreed, we have no idea at present how to live our lives without *X,* but having seen that *X* was not inevitable, in the present state of things, we can at least modify some aspects of *X,* in order to make *X* less of a bad thing. This is *reformist* constructionism. Reformist constructionism about *X,* like every kind of constructionism, starts from (0).

On the other side of irony is what Karl Mannheim (1925/1952, 140) called "the unmasking turn of mind," which does not seek to refute ideas but to undermine them by exposing the function they serve. Mannheim had learned from Marxism. The notion is that once one sees the *"extra-theoretical function"* (Mannheim's emphasis) of an idea, it will lose its "practical effectiveness." We unmask an idea not so much to "disintegrate" it as to strip it of a false appeal or authority. This is *unmasking* constructionism. A reformist may be an unmasker, or may not be; an unmasker may or may not be reformist. That is why, in my little table, I place the two grades of commitment side by side.

Unmaskers, at least as understood by Mannheim, believe not only (1) that *X* is not inevitable, but also (2) that *X* is a bad thing, and probably (3) that we would be better off without *X.* Unmasking is nevertheless an intellectual exercise in itself. A great deal of gender politics goes further, and is unequivocally radical about (1), (2), and (3), so far as concerns gender relations. A constructionist who actively maintains (1), (2), and (3) about *X* will be called *rebellious* about *X.* An activist who moves beyond the world of ideas and tries to change the world in respect of *X* is *revolutionary.*

As our consciousness about gender is raised, some of us find our attitudes moving along from historical to ironic to reformist, and then to unmasking the function of gender relations. With the mask removed, we become rebellious; a few become revolutionary.

Recall the economy. How could we possibly think about the industrial world without thinking about the economy? That is where our ironic, perhaps unmasking, social constructionist could enter. The ironist shows how the idea of the economy became so entrenched; it did not have to be, but now it is so much a part of our way of thinking, we cannot escape it. The unmasker exposes the ideologies that underlie the idea of the economy and shows what extra-theoretical functions and interests it serves. In former times there were activists who would have passed on to rebellion and even revolution about the idea of the economy. Their task becomes harder and harder with the hegemony of the world system. What once was visibly contingent feels like it has become part of the human mind. It takes only a little fortitude to be a rebellious constructionist about the idea of the deficit. But perhaps the only way you can begin to be a constructionist about the idea of the economy is to pass at once from irony to revolution.

OBJECTS, IDEAS, AND ELEVATOR WORDS

Three distinguishable types of things are said to be socially constructed. The resulting divisions are so general and so fuzzy at the edges that felicitous names do not come to hand. In addition to "objects" and "ideas" we need to take note of a group of words that arise by what Quine calls semantic ascent: truth, facts, reality. Since there is no common way of grouping these words, I call them elevator words, for in philosophical discussions they raise the level of discourse.

Objects. Items in the following disparate list are "in the world" in a commonsensical, not fancy, meaning of that phrase.

People (children)
States (childhood)
Conditions (health, childhood autism)
Practices (child abuse, hiking)
Actions (throwing a ball, rape)
Behavior (generous, fidgety)
Classes (middle)

Experiences (of falling in love, of being disabled)
Relations (gender)
Material objects (rocks)
Substances (sulphur, dolomite)
Unobservables (genes, sulphate ions)
Fundamental particles (quarks)

And homes, landlords, housecleaning, rent, dry rot, evictions, bailiffs, squatting, greed, and the Caspian Sea. The id is an object, if there is an id, and who doubts that there are egos, big ones, in the world? These items of very different categories are all in the world, so I call them objects, for lack of a better label. Adapting a terminology of John Searle's (1995), we find that some of these items are ontologically subjective but epistemologically objective items. The rent you have to pay is all too objective (and in the world, as I put it) but requires human practices in order to exist. It is ontologically subjective, because without human subjects and their institutions there would be no such object as rent. But rent is epistemologically objective. You know full well (there is nothing subjective about it) that $850 is due on the first of the month.

Ideas. I mean ideas, conceptions, concepts, beliefs, attitudes to, theories. They need not be private, the ideas of this or that person. Ideas are discussed, accepted, shared, stated, worked out, clarified, contested. They may be woolly, suggestive, profound, stupid, useful, clear, or distinct. For present purposes, groupings, classifications (ways of classifying), and kinds (the woman refugee) will be filed as ideas. Their extensions—classes, sets, and groups (the group of women refugees now meeting with the Minister of Immigration)—are collections in the world, and so count as "objects." I am well aware that there is much slippage in this coarse system of sorting.[12]

Elevator words. Among the items said by some to be constructed are facts, truth, reality, and knowledge. In philosophical discussions, these words are often made to work at a different level than words for ideas or words for objects, so I call them elevator words. Facts, truths, reality, and even knowledge are not objects in the world, like periods of time, little children, fidgety behavior, or loving-kindness. The words are used to say something about the world, or about what we say or think about the world.[13] They are at a higher level. Yes, there is a correspondence theory of truth, according to which true propositions correspond to facts. So are not facts "in the world"? They are not in the world in the same way that homes, greed, and bailiffs are in the world. Even if we agreed with Wittgenstein that the world is made up of facts and not things, facts would not be *in* the world, in the way in which greed and bailiffs are.

There are two particular points to note about elevator words. First, they tend to be circularly defined. Compare some desk dictionaries. One would hardly know that the word "fact," as defined in *Webster's New Collegiate,* is the same word as that defined in *Collins.* The *American Heritage Dictionary* begins with "1. Information presented as objectively real." It plays it safe with those two words at the end, but blows it with "presented"—you mean something could be a fact just because it is *presented* as objectively real? The *New Shorter Oxford* gives as one sense of "real," "that is actually and truly such." J. L. Austin and his fellow 1950s philosophers of language are said to have played a game called *Vish!* You look up a word, and then look up words in its dictionary definition; when you have got back to the original word, you cry *Vish!* (vicious circle). Try that on the *New Shorter Oxford* entries for "real" and break some records.

A second point to notice is that these words, along with their adjectives such as "objective," "ideological," "factual," and "real" (not to mention the "objectively real" of the *American Heritage*); have undergone substantial mutations of sense and value (Daston 1992, Daston and Galison 1992, Shapin 1995, Poovey 1998). Some of the most general, and venomous, debates about social construction end up with arguments heavily loaded with these words, as if their meanings were stable and transparent. But when we investigate their uses over time, we find that they have been remarkably free-floating. This is not the place to explore such issues. The difficulties with these nouns and adjectives provide one reason for being wary of arguments in which they are used, especially when we are asked to glide from one to the other without noticing how thin is the ice over which we are skating.

Despite these difficulties, we can agree that a thesis about the construction of a fact is different in character from a thesis about the construction of the child viewer of television, for it is not about the construction of either an object or an idea. One place we encounter the alleged construction of facts is in the sciences, as in the subtitle of Latour and Woolgar's (1986) *Laboratory Life: The Construction of Scientific Facts* (see Chapter 3). What about the social construction of reality? That sounds like the social construction of everything.

UNIVERSAL CONSTRUCTIONISM

The notion that everything is socially constructed has been going the rounds. John Searle (1995) argues vehemently (and in my opinion cogently) against universal constructionism. Yet he does not name a single universal constructionist. Sally Haslanger (1995, 128) writes that "On occasion it is possible to find the claim that 'everything' is socially constructed 'all the way down.' " She cites only a single allusive pair of pages out of the whole of late twentieth-century writing (namely Fraser 1989, pages 3 and 59, writing about Foucault), as if she had a hard time finding even one consistently self-declared universal social constructionist.

We require someone who claims that every object whatsoever—the earth, your feet, quarks, the aroma of coffee, grief, polar bears in the Arctic—is in some nontrivial sense socially constructed. Not just our experience of them, our classifications of them, our interests in them, but these things themselves. Universal social constructionism is descended from the doctrine that I once named linguistic idealism and attributed, only half in jest, to Richard Nixon (Hacking, 1975, 182). Linguistic idealism is the doctrine that only what is talked about exists; nothing has reality until it is spoken of, or written about. This extravagant notion is descended from Berkeley's idea-ism, which we call idealism: the doctrine that all that exists is mental.

Universal social constructionism is in this vein of thought, but it has not yet found its Berkeley to expound it. Most constructionism is not universal. The authors who contributed books for my alphabetical list of topics, from authorship to Zulu nationalism, were making specific and local claims. What would be the point of arguing that danger, or the woman refugee, is socially constructed, if you thought that everything is socially constructed?

But is there not an obvious example of universal constructionism, even in my alphabetical list? I mean *R* for Reality. The very first book to have "social construction" in the title was by Peter Berger and Thomas Luckmann (1966): *The Social Construction of Reality.* They argued that our experience of reality, our sense of reality as other, in all its rich and circumstantial detail, as independent of us, is neither a Kantian *a priori* nor solely the product of psychological maturation. It is the result of processes and activities which they thought might aptly be called social construction. Their book has roots in phenomenology, and especially the 1930s work of the Viennese social theorist Alfred Schutz (1899–1959). Schutz worked at the New School for Social Research after 1939. His philosophical roots were in Edmund Husserl and Max Weber. Where Husserl had asked us, in his middle years, to reflect on the quality of immediate experiences, and Weber had directed us to the fabric of society as a way to understand ourselves and others, Schutz brought the two together. His project was to understand the taken-for-granted and experienced world that each person in a society shares with others. That is the topic for Berger and Luckmann, themselves closely associated with Frankfurt and with the New School.

Their book, then, is about the social construction of our sense of, feel for, experience of, and confidence in, commonsense reality. Or rather, as the authors made plain from the start, of various realities that arise in the complex social worlds we inhabit. The book thus contrasts with psychological accounts of the origins of our conceptions of space, number, reality, and the like advanced by Jean Piaget and his colleagues. According to Berger and Luckmann, the experience of the world as other is constituted for each of us in social settings. The two authors began by examining what they called "everyday reality," which is permeated by both social relations and material objects. They moved at once to what they said is the prototypical case of social interaction, "the face-to-face situation," from which all other cases are, they held, derivative.

Berger and Luckmann did not stake a claim for any form of universal social constructionism. They did not claim that everything is a social construct, including, say, the taste of honey and the planet Mars—the very taste and planet themselves, as opposed to their meanings, our experience of them, or the sensibilities that they arouse in us. As their subtitle said, they wrote *A Treatise in the Sociology of Knowledge.* They did not claim that nothing can exist unless it is socially constructed.

THE CHILD VIEWER OF TELEVISION

As you run down my alphabetical list, you seem to see what I call objects, and a few elevator words, but no ideas. Yet that is misleading, for on closer inspection, it seems to be the idea of danger, or the classification of individuals as women refugees, that is being discussed. One of the first social-construction-of books to be published after Berger and Luckmann was Jack Douglas's (1970) *Deviance and Respectability: The Social Construction of Moral Meanings.* That makes it nicely clear that meanings, not deviance and respectability themselves, are the primary focus of discussion. Of course deviance and respectability themselves are formed in social settings, but that is not the topic of this intelligent book by the author of a famous work on suicide. Much later there is a treatment of the subject with a less clear title, *The Social Construction of Deviance* (Goode 1994).

The most banal example on my list is the child viewer. It is urged that the very idea of this definite kind of person, the child viewer of television, is a construct. Although children have watched television since the advent of the box, there is (it is claimed) no definite class of children who are "child viewers of television" until "the child viewer of television" becomes thought of as a social problem. The child viewer, steeped in visions of violence, primed for the role of consumer, idled away from healthy sport and education, becomes an object of research. Putting it crudely, what is socially constructed, in this case, is an *idea,* the idea of the child viewer. Once again "the whale" comes to mind; "the child viewer" becomes a species of person. The idea works. V-chips are invented in a Vancouver basement, devices to allow children to watch only the shows favored by parental guidance (or Parental Guidance), chips that are then to be embedded in TV sets, while talk about chips becomes part of the rhetoric of a United States presidential campaign.

The story continues. At one point when I was thinking about social constructs, there was a world congress on the child viewer of television.[14] Previously research had been conducted only in advanced industrial countries, and chiefly in English. In 1997, researchers from Chile and Tunisia could have their say alongside their well-established colleagues. Certain absences were conspicuous: children, producers, advertisers, products, and television sets as objects of study (as opposed to mere devices for use at the conference). Nevertheless, The Child Viewer advanced. No longer passive victims, children were presented as active, as masters of the screen, controllers of their world, or at any rate participants alongside the image-makers.

We have presupposition (0): The child viewer seems like an inevitable categorization in our day and age. The constructionist argues (1): Not at all. Children who watch television need never have been conceptualized as a distinct kind of human being. What seems like a sensible classification to use when thinking about the activities of children, has, it may be argued, been foisted upon us, in part because of certain moralizing interests. Hence there is also a strong implication of (2), that this category is not an especially good one. Perhaps also a suggestion of (3), that we would be better off without it. Talk about the child viewer is not exactly false, but it uses an inapt idea. It presupposes that there is a coherent object, the child viewer of television. Yes, we can collect data about watching television, ages, sex, parental status, shows, duration, attentiveness, school scores. These are not, however, very meaningful data: they are artifacts of a construction that we would be better off without, or so says the unmasker.

Once we have the phrase, the label, we get the notion that there is a definite kind of person, the child viewer, a species. This *kind* of person becomes reified. Some parents start to think of their children as child viewers, a special type of child (not just their kid who watches television). They start to interact, on occasion, with their children regarded not as their children but as child viewers. Since children are such self-aware creatures, they may become not only children who watch television, but, in their own self-consciousness, child viewers. They are well aware of theories about the child viewer and adapt to, react against, or reject them. Studies of the child viewer of television may have to be revised, because the objects of study, the human beings studied, have changed. That species, the Child Viewer, is not what it was, a collection of some children who watch television, but a collection that includes self-conscious child viewers.

Thus a social construction claim becomes complex. What is constructed is not only a certain classification, a certain kind of person, the child viewer. It is also children who, it might be argued, become socially constructed or reconstructed within the matrix. One of the reasons that social construction theses are so hard to nail down is that, in the phrase "the social construction of *X*," the *X* may implicitly refer to entities of different types, and the social construction may in part involve interaction between entities of the different types. In my example, the first reference of the *X* is a certain classification, or kind of person, the child viewer. A subsidiary reference may be children themselves, individual human beings. And yet not simply the children, but their ways of being children, Catherine-as-a-child-viewer-of-television. So you see that "the social construction of what?" need not have a single answer. That causes a lot of problems in constructionist debates, People talk at cross purposes because they have different "whats" in mind. Yet it is precisely the interaction between different "whats" that makes the topic interesting.

And confusing, for there are lots of interactions. Consider one reason that the scholars at the 1997 World Congress on the child viewer suddenly acknowledged that children are not passive victims. It is because new technologies have made children interact with screens. Not just middle-class children with family PCs, but the poor in video arcades. Children's relationships to screens change because of changes in the material world of manufacture and commerce. But they also change because of the way in which these phenomena are conceptualized.

There are many examples of this multi-leveled reference of the *X* in "the social construction of *X*." It is plain in the case of gender. What is constructed? The idea of gendered human beings (an idea), and gendered human beings themselves (people); language; institutions; bodies. Above all, "the experiences of being female." One great interest of gender studies is less how any one of these types of entity was constructed than how the constructions intertwine and interact, how people who have certain "essential" gender traits are the product of certain gendering institutions, language, practices, and how this determines their experiences of self.

In the case of the child viewer I may have stretched things to find more than one reference for the *X*; in the case of gender there are allusions to a great many different *X*s. What about the construction of Homosexual Culture? Are we being told about how the idea of there being such a culture, was constructed, or are we being told that the culture itself was constructed? In this case a social construction thesis will refer to both the idea of the culture and to the culture, if only because some idea of homosexual culture is at present part of homosexual culture.

WHY WHAT? FIRST SINNER, MYSELF

Why bother to distinguish ideas from objects, especially if many writers use one word, *X*, to refer to both objects of a certain sort and the sort itself, the idea under which the objects are thought about? Because idea and object are often confused. I have done it myself.

In *Rewriting the Soul* (Hacking 1995) I referred to a paper by a pediatrician titled "The Social Construction of Child Abuse" (Gelles 1975).

We have since had a book with that subtitle (Janko, 1994), and a thesis titled "The Social Construction of Child Neglect" (Marshall 1993), so this topic is still timely. In order to forestall tedious discussion about whether child abuse was socially constructed or real, I wrote that "it *is* a real evil, and it was so before the concept was constructed. It was nevertheless constructed. Neither reality nor construction should be in question" (Hacking 1995, 67f).[15]

What a terrible equivocation! What "it" is a real evil? The object, namely the behavior or practice of child abuse. What "it" is said to be socially constructed? The concept. My switch from object (child abuse) to idea (the concept of child abuse) is worse than careless. But not so fast. I thought, in retrospect, that I had been guilty of careless confusion, yet a number of people have told me how the very same passage has been helpful to them. It gave some readers a way to see that there need be no clash between construction and reality. We analytic philosophers should be humble, and acknowledge that what is confused is sometimes more useful than what has been clarified. We should diagnose this situation, and not evade it.

My diagnosis is that my error conceals the most difficult matter of all. As illustrated even by the child viewer of television, concepts, practices, and people interact with each other. Such interaction is often the very point of talk of social construction. My original plan for studying child abuse was largely motivated by an attempt to understand this type of interaction, which goes right back to my project of "making up people" (Hacking 1986). However, the fact that I was constantly aware of all that is no excuse. I still conflated two fundamentally different categories.

WHY WHAT? SECOND SINNER, STANLEY FISH

Directly after Sokal's notorious hoax and self-exposure, Fish sent an op-ed piece to *The New York Times*. He was at pains (in this respect like me, alas) to urge that something can be both socially constructed and real. Hence (urged Fish) when the social constructionists are taken to say that quarks are social constructions, that is perfectly consistent with saying that quarks are real, so why should Sokal get into a tizzy?

Fish argued his case by saying that baseball is a social construction. He took as his example balls and strikes.[16] "Are balls and strikes socially constructed?" he asked, "Yes. Are balls and strikes real? Yes." Fish may have meant to say that the idea of what a strike is, is a social product. If he had used Searle's terminology, he might have said that strikes are epistemologically objective: whether or not someone struck out is an objective fact. ("Kill the ump!" you cry, because you think the umpire made an objectively wrong decision.) But strikes are ontologically subjective. There would be no strikes without the institution of baseball, without the rules and practices of people.

Fish wanted to aid his allies, but did nothing but harm. Balls and strikes are real *and* socially constructed, he wrote. Analogously, he was arguing, quarks are real *and* socially constructed. So what are Sokal and company so upset about? Unfortunately for Fish, the situation with quarks is fundamentally different from that for strikes. Strikes are quite self-evidently ontologically subjective. Without human rules and practices, no balls, no strikes, no errors. Quarks are not self-evidently ontologically subjective. The shortlived quarks (if there are any) are all over the place, quite independently of any human rules or institutions. Someone may be a universal constructionist, in which case quarks, strikes, and all things are socially constructed, but you cannot just say "quarks are like strikes, both real and constructed." How might Fish have argued his case?

Perhaps it is the idea of quarks, rather than quarks, which is the social construction. Both the process of discovering quarks and the product, the concept of the quark and its physical applications, interest historians of science. Likewise for ideas of, and the theory behind, Maxwell's Equations, the Second Law of Thermodynamics, the velocity of light, and the classification of dolomite as a significant variant of limestone. All these ideas have histories, as does any idea, and they have different types of history, including social histories. But quarks, the objects themselves, are not constructs, are not social, and are not historical.

I am taking some liberties here, which I will correct in Chapter 3. Andrew Pickering's *Constructing Quarks* (1986) is the only systematic social construction work about quarks. I would trivialize its central themes if I tried to turn it into a mere social and material history of the idea of the quark. Not surprisingly, Pickering wrote, in a letter of 6 June 1997: "I would never say that *Constructing Quarks* is about 'the idea of quarks.' That may be your take on constructionism re the natural sciences, but it is not mine. My idea is that if one comes at the world in a certain way—your heterogeneous matrix—one can elicit certain phenomena that can be construed as evidence for quarks."

The problem with that final sentence is, who would disagree with it? Pickering's interesting claim is a converse of what he wrote: if you came at the world in another way, you could elicit other phenomena that could be construed as evidence for a different (not formally incompatible, but different) successful physics. Pickering holds that the evolution of physics, including the quark idea, is thoroughly contingent and could have evolved in other ways, although subject to very different types of resistance than, say, the conservatism of ballplayers.

Most physicists, in contrast, think that the quark solution was inevitable. They are pretty sure that longstanding parts of physics were inevitable. There is a significant point at issue here, which Fish's inept conciliation conceals. In Chapter 3 I call this disagreement about contingency "sticking point #1" in the science wars. Far from wanting to sweep it under the carpet, I want to make it a central piece of furniture in the parlor of debate. Unlike Stanley Fish, I do not want peace between constructionist and scientist. I want a better understanding of how they disagree, and why, perhaps, the twain shall never meet.

INTERACTIONS

We have seen how some objects and ideas may interact. The idea of the child viewer of television interacts with the child viewer. Ways of classifying human beings interact with the human beings who are classified. There are all sorts of reasons for this. People think of themselves as of a kind, perhaps, or reject the classification. All our acts are under descriptions, and the acts that are open to us depend, in a purely formal way, on the descriptions available to us. Moreover, classifications do not exist only in the empty space of language but in institutions, practices, material interactions with things and other people. The woman refugee—that kind or "species" of person, not the person—is not only a kind of person. It is a legal entity, and more importantly a paralegal one, used by boards, schools, social workers, activists—and refugees. Only within such a matrix could there be serious interaction between the "kind" of person and people who may be of that kind.

Interactions do not just happen. They happen within matrices, which include many obvious social elements and many obvious material ones. Nevertheless, a first and simplistic observation seems uncontroversial. It stems from the almost-too-boring-to-state fact that people are aware of what is said about them, thought about them, done to them. They think about and conceptualize themselves. Inanimate things are, by definition, not aware of themselves in the same way. Take the extremes, women refugees and quarks. A woman refugee may learn that she is a certain kind of person and act accordingly. Quarks do not learn that they are a certain kind of entity and act accordingly. But I do not want to overemphasize the awareness of an individual. Women refugees who do not speak one word of English may still, as part of a group, acquire the characteristics of women refugees precisely because they are so classified.

The "woman refugee" (as a kind of classification) can be called an *interactive kind* because it interacts with things of that kind, namely people, including individual women refugees, who can become aware of how they are classified and modify their behavior accordingly. Quarks in contrast do not form an interactive kind; the idea of the quark does not interact with quarks. Quarks are not aware that they are quarks and are not altered simply by being classified as quarks. There are plenty of questions about this distinction, but it is basic. Some version of it forms a fundamental difference between the natural and the social sciences. The classifications of the social sciences are interactive. The classifications and concepts of the natural sciences are not. In the social sciences there are conscious interactions between kind and person. There are no interactions of the same type in the natural sciences. It is not surprising that the ways in which constructionist issues arise in the natural sciences differ from questions about construction in human affairs. I shall now pose two separate groups of questions: (1) those involving contingency, metaphysics, and stability; and (2) issues that are biological but still of the interactive kind.

TWO QUESTION AREAS

The history of science tells of definite bench marks, established facts, discovered objects, secure laws, on the basis of which subsequent inquiry proceeds, at least for some substantial period of time. Physics establishes, with Rutherford, that the atom can be split; on we go, through quantum electrodynamics, weak neutral currents, gauge theory, quarks. The Higgs boson and the lepto-quark lurk tantalizingly in the future, one predicted by theory, the other a refutation of it.

A social construction thesis for the natural sciences would hold that, in a thoroughly nontrivial sense, a successful science did not have to develop in the way it did, but could have had different successes evolving in other ways that do not converge on the route that was in fact taken. Neither a prior set of bench marks nor the world itself determines what will be the next set of bench marks in high-energy physics or any other field of inquiry. I myself find this idea hard to state, let alone to believe. One question, worthy of discussion, is how should we state the idea implicit in Pickering's work, in order to make it at least intelligible to those who are skeptical of it? Then comes the question of whether it is a good idea, a true idea, a plausible idea, a useful perspective.

If contingency is the first sticking point, the second one is more metaphysical. Constructionists tend to maintain that classifications are not determined by how the world is, but are convenient ways in which to represent it. They maintain that the world does not come quietly wrapped up in facts. Facts are the consequences of ways in which we represent the world. The constructionist vision here is splendidly old-fashioned. It is a species of nominalism. It is countered by a strong sense that the world has an inherent structure that we discover.

The third sticking point is the question of stability. Contrary to the themes of Karl Popper and Thomas Kuhn, namely refutation and revolution, a great deal of modern science is stable. Maxwell's Equations, the Second Law of Thermodynamics, the velocity of light, and lowly substances such as dolomite are here to stay. Scientists think that the stability is the consequence of compelling evidence. Constructionists think that stability results from factors external to the overt content of the science. This makes for the third sticking point, internal versus external explanations of stability.

Each of these three sticking points is the basis of genuine and fundamental disagreement. Each is logically independent of the others. Moreover, each can be stated without using elevator words like "fact," "truth," or "reality," and without closely connected notions such as "objectivity" or "relativism." Let us try to stay as far as we can from those blunted lances with which philosophical mobs charge each other in the eternal jousting of ideas.

A second group of questions arises in human affairs rather than in the theoretical and experimental natural sciences. We have seen that very commonly, when people talk of the social construction of *X*, they have in mind several interacting items, all designated by *X*.

To return to my alphabetical list, many of the items, such as authorship or brotherhood, are built around kinds of people such as authors and brothers (in the sense of solidarity, not blood). *Author* and *brother* are kinds of people, as are *child viewer* and *Zulu*. People of these kinds can become aware that they are classified as such. They can make tacit or even explicit choices, adapt or adopt ways of living so as to fit or get away from the very classification that may be applied to them. These very choices, adaptations or adoptions have consequences for the very group, for the kind of people that is invoked. The result may be particularly strong interactions. What was known about people of a kind may become false because people of that kind have changed in virtue of what they believe about themselves. I have called this phenomenon *the looping effect of human kinds* (Hacking 1995).

Looping effects are everywhere. Think what the category of genius did to those Romantics who saw themselves as geniuses, and what their behavior did in turn to the category of genius itself. Think about the transformations effected by the notions of fat, overweight, anorexic. If someone talks about the social construction of genius or anorexia, they are likely talking about the idea, the individuals falling under the idea, the interaction between the idea and the people, and the manifold of social practices and institutions that these interactions involve: the matrix, in short.

Chapter Two

TOO MANY METAPHORS

The metaphor of social construction once had excellent shock value, but now it has become tired. It can still be liberating suddenly to realize that something is constructed and is not part of the nature of things, of people, or human society. But construction analyses run on apace.[1]

Looking at their many titles makes one wonder what work the phrase "social construction" is doing. Take the entry for *L: The Social Construction of Literacy* (Cook-Gumperz 1986). The editor begins with an article of her own with the same title. There is no indication of what "social construction" means, nor any attempt to exemplify it. The book is about innovative ways of teaching children to read. The children are often disadvantaged; then they learn to read, both in and out of the California school system. Now it certainly is possible to think of literacy—the idea of literacy—as a social construct, with a good many political overtones (Hacking 1999). But that was not the point of the book at hand. It undertakes the valuable task of presenting a "social perspective" on how children learn to read, or don't. Why talk of social construction? We fear a case of bandwagon-jumping.

Construction has been trendy. So many types of analyses invoke social construction that quite distinct objectives get run together. An all-encompassing constructionist approach has become rather dull—in both senses of that word, boring and blunted. One of the attractions of "construction" has been the association with radical political attitudes, stretching from bemused irony and angry unmasking up to reform, rebellion, and revolution. The use of the word declares what side one is on.

Sometimes this declaration tends to complacency. Sometimes uttering the very phrase "social construction" seems more like standing up at a revival meeting than enunciating a thesis or project. Two things are readily forgotten. One is that a great many social construction discussions are embedded in the conception of a social problem that began, for American professors, perhaps a century ago. It led in due course to the journal *Social Problems,* and a gifted set of sociologists centered in Chicago. The trouble is that social construction has become a part of the very discourse that it presents itself as trying to undo.[2]

Secondly, it is astonishingly easy to lose the whole picture while focusing on a single pixel. Some constructionists wish to declare a kind of ownership over the context in which a social problem emerged, with the view that the outrages of times gone by are the same outrages which determine the present. This antiquarian view exists as a veneration for the past—though a strange veneration, which its practitioners would be insulted to hear so described. Such a position may suffer from myopia, for "most of what exists it does not see at all, and the little it does see it sees much too close up and isolated; it cannot relate what it sees to anything else and it therefore accords everything it sees equal importance and therefore to each individual thing too great importance" (Nietzsche 1874/1983, 74).

PROCESS AND PRODUCT

Most words ending in "tion" are ambiguous between process and product, between the way one gets there, and the result. The termination of the contract: that can mean the process of terminating the contract. It can also mean the upshot, the product, the end of the contract. The pattern is not identical for each "tion" word, because each word nuances the ambiguity in its own way. "Production" itself can mean the process of producing, or, in other circumstances, the result of producing. Is the production of a play process or product? What about movies?

As Lewis White Beck (1950, 27) noted long ago, our word "construction" shares in this ambivalent pattern of ambiguity. Thus we read, in a travel guide to Japan, that the construction of the Garden of Katsura Rikyu, the Imperial Villa in Kyoto, was completed by Toshihito in 1620. This refers to a process that came to an end in 1620. Then we read that the garden is a remarkably meaningful formal construction which consists of a semiformal pavement combining cut and irregular stones, followed by a series of natural stepping stones, called jumping stones, which contrast with the stylized cut stones of the villa at the end of the path. In this sense it is the product that is meaningful, a delicate play between art and nature, that might not even have been intended by Toshihito.

Construction-as-process takes place in time. Some social construction books make this plain in a subtitle. Pickering on constructing quarks (1986) is subtitled *A Sociological History of Particle Physics.* Danziger (1990) on constructing the subject is subtitled *Historical Origins of Psychological Research.* The recourse to history is implied by other phrases, as in "The invention of teenage pregnancy" (Arney and Bergen 1984), or "The 'making' of teenage pregnancy" (Wong 1997). When Latour and Woolgar (1979) wrote of the construction of a scientific fact, they wrote a fragment of the history of endocrinology. It is true that Latour presented himself as an anthropologist, and many others who write about the sciences present themselves as sociologists. Nevertheless their individual case studies are histories. The waters may seem a little muddied here. Some of the most prominent early social studies of science came from Edinburgh in the 1970s. The Edinburgh school, as it was called, identified its work as sociological, and claimed that it was engaged in a scientific study of science. The theoretical positions of leading figures such as David Bloor and Barry Barnes, updated in Barnes, Bloor, and Henry (1996), were more the result of philosophy than sociology. The empirical work done by the school, well represented also by MacKenzie (1981) or Mulkay (1979), was historical in character. Trained historians would often write differently about the phenomenon of teenage pregnancy than do Wong (a philosopher), or Arney and Bergen (sociologists), but the description and analysis of the process of construction, in all these cases, are historical in character.

Construction stories are histories, but to insist on only that angle is to miss the point. Constructionists about *X* usually hold that *X* need not have existed, or need not be at all like it is. Some urge that *X* is quite bad, as it is, and even that we would be much better if *X* were done away with, or at least radically transformed. *X,* the product, is the focus of attention, although, as I have explained, it is usually not *X,* the thing, teenage pregnancy, but the idea of *X,* the idea of teenage pregnancy, and the matrices in which the idea has life. If we overhear someone say, "And even I am a social construction," we know that it is the person as product who is in question, a person who has been constructed by a social process, that person's life history.

Process and product are both part of arguments about construction. The constructionist argues that the product is not inevitable by showing how it came into being (historical process), and noting the purely contingent historical determinants of that process.

In the next chapter I turn to natural sciences such as physics, chemistry, and molecular biology. Social construction provides one arena for the science wars. Constructionists state that various items from the natural sciences are social constructs. Many scientists deny that. They will admit that there is a (social) history of the discovery of the item in question, say the Second Law of Thermodynamics. Once upon a time the Second Law had ideological, political, or religious overtones. That does not matter. "The Second Law of Thermodynamics is neither an empirical claim, nor a social construction, nor a consensus by institutionalized experience, but an inexorable law based on the atomic constitution of matter" (Perutz 1996, 69). It is a fact about the universe that we have discovered. The history of its discovery makes no jot of difference to what it is, was, and always will be.

Disability

In social affairs, as opposed to chemistry or physics, scholars do make distinguishable claims, some meaning process and some meaning product. Take discussions of disability. We read that "disability as a category can only be understood within a framework which suggests that it is culturally produced and socially structured" (Oliver 1984, 15).[3] The "it" that is "culturally produced" is a product. The cultural production is process. The "socially structured" is ambiguous. It could mean that the product is socially structured, in the sense that it has a structure that exists in a social setting (a structure reminiscent of the synchronic structure of Parisian structuralism). Or it could mean that the product is organized by a historical process named social structuring.

Sometimes process is clearly intended. "I call the interaction of the biological and the social to create (or prevent) disability the 'social construction of disability.' " (Wendell 1993, 22). Now examine this statement: "The disabled individual is an ideological construction related to the core ideology of individualism" (Asche and Fine 1988, 13). "The disabled individual" may refer either to a kind of person, almost a subspecies (as in "the whale is . . .") or to individuals of that kind, particular disabled individuals. In either case the author refers to a product, the kind or the individuals.

There is yet another sense of construction, in addition to product and process. It has the same etymological roots as, and is similar in meaning to, "construal." "Construal" originally meant seeing how a sentence is to be understood on the basis of its component parts. But the word quickly acquired the sense of interpretation. In the United States, a strict constructionist is a constitutional expert who argues for a strict construal of the American Constitution, trying to go no further than the very words written down and agreed upon by the founding fathers.

Harlan Lane is a distinguished deaf rights advocate, and partisan of American Sign Language as the basis for a viable linguistic community. He wrote an essay titled "The Social Construction of Deafness." There he mentioned two "constructions of deafness, which are dominant and compete for people's destinies" (Lane 1975, 12). What he meant was two ways of understanding deafness, two ways of thinking about deafness, two ways of construing deafness. One way to construe deafness is to think of it as a disability. Another way to construe deafness is to think of it as the basis for the formation of a linguistic minority. Construal, construction-as-process, and construction-as-product are inevitably intertwined, but to fail to distinguish them is to fall victim to forgotten etymologies.

IS "SOCIAL" REDUNDANT?

Most items said to be socially constructed could be constructed only socially, if they are constructed at all. Hence the epithet "social" is usually unnecessary and should be used sparingly, and only for emphasis or contrast. Take for example the *G* entry for my alphabetical list in Chapter 1. Lorber and Farrell's anthology (1991) is entitled *The Social Construction of Gender.* I have already sketched a diversity of feminist approaches to gender. Yet no matter what definition is preferred, the word is used for distinctions among people that are grounded in cultural practices, not biology. If gender is, by definition, something essentially social, and if it is constructed, how could its construction be other than social? The point seems to become self-evident when we get to titles like *The Social Construction of Social Policy* (Samson and Smith 1996).

The emphasis made with the word "social" becomes useful when we turn to inanimate objects, phenomena, or facts that are usually thought of as part of nature, existing independent of human society. This is true for Latour and Woolgar is (1979) book, subtitled *The Social Construction of Scientific Facts.* They described work done in a laboratory whose head shared a Nobel prize for medicine for discovering the structure of a certain tripeptide, a hormone called Thyrotropin Releasing Hormone. What, according to the authors, was socially constructed? The fact, they answer, that this hormone was such and such a tripeptide. The hormone, and the new methods for establishing its structure, were thought to be so important that they earned the Nobel prize. So it was shocking, in 1979, to be told about the social construction of such an impersonal, presocial, biochemical fact. Yet in their second edition, Latour and Woolgar (1986) dropped the word "social" from their subtitle: "What does it mean to talk about 'social' construction? There is no shame in admitting that the term no longer has any meaning . . . By demonstrating its pervasive applicability, the social study of science has rendered 'social' devoid of any meaning" (p. 281). Latour had his own agenda here, increasingly apparent later with the "hybrid natural/social actants" (Latour 1987) and the "parliament of things" (Latour 1993). He holds that the usual distinction between the natural and the social is a sham. But one need not agree with his agenda in urging that we drop the "social," except for an occasional emphasis.

Now turn to essentially social entities, states, or conditions—I strive for sufficiently generic and noncommittal nouns here—such as literacy or lesbianism. If literacy is constructed, how other than socially? Perhaps being lesbian is an innate characteristic of some women, but if lesbianism is constructed, how other than socially? The philosopher-sociologists of the natural sciences seem to have been ahead of those who study more humane topics such as lesbianism or literacy. They banned the adjective "social" from their titles and their texts. Authors discussing specifically human affairs continued to employ it rather unreflectively.

KANT'S HOUSE

It is not always pointless to use the word "social" in connection with construction. For example, "social constructionist" has come to name a quite widespread body of tenets, theories, or attitudes. The adjective "social" is part of the name of this body of thought. Thus Donna Haraway (1991, 184) wrote that "recent social studies of science and technology have made available a very strong social constructionist argument for all forms of knowledge claims, most certainly and especially scientific ones." She cited Knorr-Cetina and Mulkay (1983), Bijker et al (1987), and "especially Latour" (1987) on Pasteur. Although Latour would erase the adjective "social," it is useful for Haraway to have a name for the school of constructionism that she takes to be represented by Latour, Knorr-Cetina, Mulkay, and Bijker. This is because there are many other schools. All of them, including social constructionism, seem to derive from Kant.

Kant was the great pioneer of construction. Onora O'Neill's book about Kant, *Constructions of Reason* (1989), is well titled. Kant was truly radical in his day, but he still worked within the realm of reason, even if his very own work signaled the end of the Enlightenment. After his time, the metaphor of construction has served to express many different kinds of radical philosophical theory, not all of them dedicated to reason. But all agree with Kant in one respect. Construction brings with it one or another critical idea, be it the criticism of the *Critique of Pure Reason* or the cultural criticisms advanced by constructionists of various stripes. We have logical constructions, constructivism in mathematics, and, following Kant, numerous strains of constructionism in ethical theory, including those of John Rawls and Michel Foucault.

Bertrand Russell's Logical Constructions

"Wherever possible," wrote Russell, "logical constructions are to be substituted for inferred entities" (Russell 1918, 155). When you infer an entity, you infer that it exists. Do numbers exist? Do electrons exist? We infer (thought Russell) that electrons exist from the reliability of scientific laws involving electrons. Platonists suppose that numbers exist. Russell urged ontological caution. He did not like us to infer the existence of things of some kind, unless we could be certain that things of that kind do exist. Yet he did not want to follow the skeptic Diogenes to the bathtub, feigning ignorance about everything. We know a good deal that we express in terms of numbers and electrons. Russell wanted to be able to state what we do know, without assuming the existence of such things. That is where the notion of a logical construction comes in. On the surface, we appear to be talking about things of a certain kind, but when we analyze more deeply, we are not.

More technically: Let *T* be a term that, grammatically, is used to refer to *X*, either an individual thing, or things of a certain kind. *T* is shown to be a name for a logical construction when sentences in which *T* appears are, in a systematic way, logically equivalent to sentences in which *T* does not occur, and no reference is made to *X*. Thus although statements using *T* appear to refer to *X*, and hence to imply or presuppose the existence of *X*, logical analysis obviates the implication.

What is the point? When an inferred entity *X* is replaced by a logical construction, statements about *X* may be asserted without implying the existence of *X*s, since the logical form or deep structure of those sentences makes no reference to *X*. We are allowed to talk about *X*s while being agnostic about the existence of *X*s. This is not a "same-level" skepticism which outright denies that we have grounds for thinking that there are *X*s. Russell's analysis shows that the logical form of statements about *X* is not what we think. We discover that below the grammatical surface we were never talking about so-and-sos in the first place. Russellian analyses do not debunk inferred entities. They show that there is no commitment to the existence of so-and-sos. But they do license statements about so-and-sos, precisely because they show that those statements do not have the existential commitments we expect them to have.

Logical Positivism

Logical positivism, usually thought of as antagonistic to constructionism, was also deeply committed to the construction metaphor. Russell's program was energetically pursued in Rudolf Carnap's *Der logische Aufbau der Welt* (1928). The English translation renders *Aufbau* as "Structure," but *Aufbau* means "construction" (or, in context, "building"), and that is what Carnap meant. He wanted to establish that the world could be built up from elements, the data of sensory experience, or perhaps items that played a role in physical science.

The logical positivists (aside from Otto Neurath) might have been troubled by some of the twists of constructionism in recent sociology. Not too upset: Thomas Kuhn is standardly presented as the originator of the modern trend toward social studies of science, but as Peter Galison (1990) has shown, there is a good deal in common between Kuhn and Carnap, and both men knew it. The roots of social constructionism are in the very logical positivism that so many present-day constructionists profess to detest.

Yet we should not overdo that statement. Kuhn said little about the social. More than once he insisted that he himself was an internalist historian of science, concerned with the interplay between ideas, not the interactions of people. His masterpiece, ever fresh, is now over thirty-five years old—truly the work of a previous generation. *The Structure of Scientific Revolutions* is rightly honored, by those who conduct social studies of the sciences, as their pre-eminent predecessor. Yet for all that Kuhn emphasized a disciplinary matrix of one hundred or so researchers, or the role of exemplars in science teaching, imitation, and practice, he had virtually nothing to say about social interaction.

Construct Validity in Empirical Psychology

Before turning to a later genre of construct-ism, another of Russell's heirs should be mentioned—this one from empirical psychology. In the late 1930s logical positivist philosophers of the natural sciences had begun using the noun "construct" for theoretical entities such as electrons (see Beck 1950 for references). It was taken up in fundamental debates in the philosophy of the social sciences, for example, in connection with historical individualism, where you find J. W. N. Watkins, a Popperian, challenged by May Brodbeck, who studied with Herbert Feigl, the distinguished logical positivist who had emigrated from Berlin to Minnesota. Watkins introduced the "anonymous individual," which Brodbeck denounced as an irreducible theoretical construct and thereby unworthy of a scientific sociology. (For a summary of 1950s debates, with references, see May 1987, 14–18.)

After World War II this usage was also transferred to the philosophy of experimental psychology (for example, MacQuorquodale and Meehl 1948). Hypothetical entities or quantities in psychology came to be called constructs. Familiar examples are IQ, or Spearman's controversial *g*, the factor called "general intelligence." How can we distinguish constructs that logical positivists took to be virtuous from those that they took to be suspicious, such as libido? When are hypothetical constructs valid? The most authoritative text on psychological testing states that "The term 'construct validity' was officially introduced into the psychometrist's lexicon in 1954 in the *Technical Recommendations for*

Psychological Tests and Diagnostic Techniques, which constituted the first edition of the 1985 *Testing Standards.* The first detailed exposition of construct validity appeared the following year in an article by Cronbach and Meehl (1955)" (Anastasi 1988, 161).

The logical positivist ancestry of construct validity has been somewhat suppressed in psychology's self-history. In 1955 Lee Cronbach (b. 1916) was rapidly establishing himself as a leading figure in education. Paul Meehl (b. 1920), one of the most sophisticated critics of much experimental and statistical psychology, was another associate of Herbert Feigl. Russell's logical constructions and Carnap's *Aufbau* were very much present at the birth of that cardinal concept of psychological testing, construct validity.

Nelson Goodman's Constructionalist Orientation

Nelson Goodman, a philosopher of both the arts and the sciences, has described his philosophical orientation as "skeptical, nominalist, and constructionalist" (Goodman 1978, 1). "Constructionalist" seems to be a word of Goodman's invention. Possibly two meanings are packed into this label. One refers to Goodman's early work. It involves making or exhibiting constructions. Goodman and Quine (1947) published "Steps towards a Constructive Nominalism," dedicated to a systematic elimination of, among other things, classes, in favor of logical constructions. Goodman's *The Structure of Appearance* (1951), based on his doctoral dissertation (1940/1990), was the heir to Carnap's *Aufbau.* His early version of constructionalism was an active philosophy which constructed, or showed how to construct.

It was also a critique of *Aufbau,* arguing that what we call the world could be constructed in many ways. Might some ways be simpler than others? No. Goodman is the author of the most trenchant of critiques of the notion that simplicity has any existence outside of the eye of the beholder. Any one world may be made in many ways, and many worlds may be made.

Goodman's philosophy evolved from Russell and Carnap. His title, *Ways of Worldmaking* (1978), means what it says. Goodman contentedly talks of making worlds, and takes for granted that it is we, people, who make them. Moreover, we do so in concert. This sounds social, but Goodman got there in a straight line from Russell and Carnap.

Goodman and his fellow constructionalists say almost nothing about actual societies or social processes. This is to some extent a generational effect. Goodman's collaborator, W. O. Quine, wrote a great deal about translation, but it tended to dwell on translation involving imaginary explorers encountering natives who live in jungles populated by fauna unknown to any real jungle, namely rabbits. Whatever be the case with Quine, whose philosophy is more given to regimentation than inquiry, Goodman's world-making has to be social: it is people who do it. Goodman has been enthusiastic about at least some social studies of construction in the natural sciences.[4] Yet his work gives no hint of any actual social process involved in world-making. Chapter 5 below starts to fill the gap with a single example; many more are needed.

Constructivism in Mathematics

Kant's house has many mansions. Kant began his *Critique of Pure Reason* by trying to understand a puzzle about the truths of arithmetic and geometry. How can we know them just by thinking, and yet apply them in the real world which exists independently of thought? The answer comes in two parts. First, all experience is in space and time, which is not a fact about experience, but a precondition for anything we call experience. Second, space is structured by the laws of geometry, and the units of time are structured by the laws of arithmetic. Both structures derive from the nature of thought itself. Thus the laws of geometry and arithmetic are *a priori,* yet anything experienced must conform to them. Hence the famous doctrine of the synthetic *a priori.* Kant's view of geometry was devastated at the beginning of this century, when it became clear through Einstein's use of Riemann's mathematics that the real world might best be described by non-Euclidean geometry: there was no pure geometry of the mind uniquely best suited for experience.

A second, arithmetical, revolution failed to take place. The *Principia Mathematica* of Whitehead and Russell (1910) was intended to undo Kant's views of arithmetic, showing that number theory could be deduced from pure logic—the numbers and all their properties were, rather literally, logical constructions. That program did not pan out, for very famous reasons, connected with the name of Kurt Gödel. And at the very same time that Whitehead and Russell were writing their opus, a rival program named intuitionism was inaugurated in Holland by L. E. J. Brouwer. The "intuition" in question had a technical connotation, alluding to what Kant called pure intuitions of space and time. According to Brouwer, number-theoretic knowledge has two sources. The first source is a rather Kantian pure intuition of number. Numbers are generated by us, as they structure the experience of counting. The second source is proof, and all that we can build up from those intuitions by proof. Proofs are generated by us, as active thinkers. That seems like a truism, but it was taken seriously by intuitionists, with remarkable consequences.

Intuitionists held that mathematical objects do not exist until they have been built up by proofs of their existence, that is, until they have been constructed by mental operations. Valid proofs must be *constructive;* that implies that a mathematical object can be assumed to exist only when, by proof, we have been able to construct it out of intuited entities. Mathematics, so often thought of as a body of eternal truths, takes place in time, and objects come into being as they are constructed. This approach has a well-known radical consequence. You cannot assume the law of the excluded middle. You cannot assume that for any proposition *p,* either *p* is true, or not-*p* is true. That is because the proposition may refer to an object that has not yet been constructed by proof. The first years of the twentieth century were revolutionary times indeed. Einstein had dethroned Kant, while Brouwer's intuitionistic reasoning challenged Aristotle. Next in line were Lenin and the new quantum mechanics, the one trying to undo capitalism and and the other undoing causality.

Brouwer's intuitionism led to various types of what are called constructive mathematics, especially constructive analysis (calculus) (Bishop 1967). As with other construct-isms, constructivism in mathematics is skeptical, because it allows us to assert the existence of objects only after we have constructed them in a sequence of mental operations. Hence it forbids us to assert the existence of many mathematical objects that most mathematicians take for granted—the continuum of real numbers, for example.

Moral Theory

I have said nothing about ethics, nor will I in these chapters. Let us record, however, that it has been a constant thrust in moral theory, from Immanuel Kant's categorical imperative to John Rawls's theory of justice and Michel Foucault's self-improvement, to insist that the demands of morality are constructed by ourselves, as moral agents, and that only those we construct are consistent with the freedom that we require as moral agents. Some readers may find it natural to couple the names of Rawls and Kant but bizarre to pair Kant and Foucault. On the contrary, Foucault began his intellectual career with Kant's *Anthropologie.* Georges Canguilhem was on the mark in calling *Les Mots et les choses* a study of the historical *a priori.* Foucault was pursuing, in his own inimitable and transformative way, Kantian ethical themes of the well-made life in his own final days.

Different names for different construct-isms

Kant may have cast the mold, but the drive for construction belongs to the twentieth century. The constructing attitude is skeptical. It is also humanist. It says that the demands of morality do not come from the idealized and not-human Father or even the idealized posthuman Son. They come from the demands on rationality that free human agents place on themselves. It says that mathematical objects are not out there in Plato's nonhuman heaven; it is we who bring them into being. It says that we should not infer the existence of minute and unobservable entities from their causal effects; instead we are to describe phenomena as they appear to us, analyzing the theoretical entities into logical constructions. It says that in experimental psychology we do not use categories found in nature but constructs whose validity is established by our practices. To cap it all, Nelson Goodman tells us about ways of world-making. Not even the world itself is safe from these philosophies of construction. It is chiefly in this company that the adjective "social" marks out a further theme. Social constructionists teach that items we had thought were inevitable are social products.

What are we to call these different mansions built within Kant's house? We can help ourselves to labels that are almost ready-made. Goodman called himself a constructionalist. So let *constructionalism* refer to the philosophical projects of Russell, Carnap, Goodman, Quine, and their associates and followers. They aim at exhibiting how, or proving that, various important entities, concepts, worlds, or whatever are constructed out of other materials.

Constructionalists may hold that constructions are made by people, together, but they do not study historical or social events or processes. Their instincts are skeptical about constructed items, and yet not profoundly so. They do not say flatly that the items do not exist, or that we cannot have grounds for believing they exist. On the contrary, we have excellent grounds, but after analysis we realize that our beliefs are not what they seem. Constructionalism is a change in the level of discourse. I see this attitude as including not only Brodbeck's critique of Watkins, but also the Cronbach and Meehl proposals—now so entrenched in the experimental psychology of measurement—for legitimating constructs in psychology which do not derive from direct observation.

Without placing any weight on the terminology, I find it convenient to leave the label *constructivism* to mathematics. That is where the term was first used, at least in modern times, and it denotes a flourishing, if minority, research activity. If we left "constructivism" to mathematics, we would avoid the confusion invited by a title such as *Social Constructivism as a Philosophy of Mathematics* (Ernest 1998), which suggests, to anyone who knows anything about mathematical constructivism, something like a social variant of Brouwer's program (a rather incoherent idea). It would have been better, I think, to speak of social constructionism as a philosophy of mathematics, a philosophy that would presumably maintain that in some sense mathematical objects, such as numbers, and mathematical facts—theorems—are social constructs. That would be analogous to constructionism about the natural sciences, although the arguments would presumably be different.

It is true that many people nowadays speak of social constructivism rather than constructionism in any context whatsoever. Throughout Chapter 1 I spoke instead of (social) constructionism. (I suspect that some readers, out of habit, actually pronounced the word as "ivism," not as "ionism.") Nothing should hang upon a spelling, or a syllable, but my usage does pay attention to the fact that recent enthusiasm for social constructs is only one mansion in Kant's big house, and it allows the others, such as mathematics, to keep the names that they chose for themselves quite a long time ago.

Hence by *constructionism* (or social constructionism if we need, on occasion, to emphasize the social) I shall mean various sociological, historical, and philosophical projects that aim at displaying or analyzing actual, historically situated, social interactions or causal routes that led to, or were involved in, the coming into being or establishing of some present entity or fact.

Most constructionists have never heard of constructivism in mathematics. Constructivists, constructionists, and constructionalists live in different intellectual milieus. Yet the themes and attitudes that characterize these isms are not so different. From all three we hear that *things are not what they seem.* All three involve iconoclastic questioning of varnished reality, of what the general run of people take for real. Surprise, surprise! All construct-isms dwell in the dichotomy between appearance and reality set up by Plato, and given a definitive form by Kant. Although social constructionists bask in the sun they call postmodernism, they are really very old-fashioned.

BUILDING, OR ASSEMBLING FROM PARTS

Construction has become stale. It can be freshened up if we insist that the metaphor retain one element of its literal meaning, that of building, or assembling from parts. After the plethora of titles cited at the start of Chapter 1, it is good to be brought back to the real world, with a book title such as *Constructing a Five String Banjo.* When it comes to banjos, we are told how to make one. Most of the (social) construction/constructing works do not exhibit anything resembling a construction. Construction has become a dead metaphor. That expression, itself a metaphor, is from Fowler's eccentric *Modern English Usage:*

> METAPHOR. 1. Live & dead m. In all discussion of m. it must be borne in mind that some metaphors are living, i.e., are offered & accepted with a consciousness of their nature as substitutes for their literal equivalents, while others are dead, i.e., have been so often used that speaker & hearer have ceased to be aware that the words used are not literal; but the line of distinction between the live & the dead is a shifting one, the dead being sometimes liable, under the stimulus of an affinity or a repulsion, to galvanic stirrings indistinguishable from life. (Fowler 1926, 348–49.)

If we are to return "construction" to life, we should attend to its ordinary meanings, as in constructing a five-string banjo. The core idea, from Latin to now, is that of building, of putting together. The fairly new (1992) *American Heritage Dictionary* first offers "to form by assembling or combining parts." Then it gives us a dead metaphor, lacking brick and mortar, or girders and concrete: "To create (an argument or a sentence, for example) by systematically arranging ideas or terms." This metaphor, like the very ancient and very dead geometrical metaphor of constructing with a ruler and compass, retains the sense of systematic arrangement of elements, which become part of a whole. Of course the whole is more than the sum of the parts, because it is a systematic arrangement, a structure. Buildings are always more than the sum of their parts.

Constructionalists (Russell) and constructivists (Brouwer) were true to the root metaphor of construction as building. Whitehead and Russell wrote down the construction of the number 1 and its successors within their system of logic. Brouwer had well-understood criteria for the building up of a mathematical object by proof. And although I would not argue the point here, it seems to me that in ethics, Kant, Rawls, and Foucault, to repeat the names of the three moralists I have mentioned, tell us how to build, and why. I urge (social) constructionists to keep the same faith. Anything worth calling a construction was or is constructed in quite definite stages, where the later stages are built upon, or out of, the product of earlier stages. Anything worth calling a construction has a history. But not just any history. It has to be a history of building. There is no harm in one person stretching a metaphor, but when many do, they kill it.

Most writers never reflect on the metaphor in "construction." Sergio Sismondo is the rare philosopher who does. He generously notices six legitimate metaphorical uses of the word in the social construct literature. In fact one of these is not a metaphor at all: scientists "construct, through material intervention, artifacts in the laboratory" (Sismondo 1996, 50) Surely a great deal of apparatus is literally, not metaphorically, built out of, or assembled, from parts? Sismondo is insightful when he includes the root philosophical metaphor of construction which, as I observed above, derives from Kant. In contrast to Sismondo, I would, however, insist that most social construct writing is almost wildly metaphorical, or rather, passes beyond metaphor. Rather than give invidious examples, it is better to mention a few authors in whose work the construction metaphor is put to good use.

The Psychological Subject

Kurt Danziger's *Constructing the Subject* (1990) is a fine example of how the construction metaphor can be used, fairly literally, when applied to a social rather than a natural science. Danziger has written a history of experimental psychology. The subject in question is none other than "the subject" who appears in the experimenters' laboratory report, once upon a time often abbreviated by the letter "*s*" to depersonalize the subject as much as possible. Today we are all subject to such tests and expect to be given them when we are growing up, are inducted into the military, try out for a corporate job, or report an inexplicable malaise to a psychiatrist.

Danziger writes about the social construction of the subject. But what is that? As is quite common in *Constructing* books, Danziger writes about constructing at least four distinct kinds of entity: a concept or idea, a practice, a body of knowledge, and individuals themselves. First, there is the idea of the subject to observe or to test in experiments. Danziger is convincing when he urges that this is not a self-evident idea that was well understood as soon as the idea of laboratory-style experimentation on the human mind came into being. The first subjects of psychological experiments were commonly the experimenters themselves—Gustav Fechner, for example. Or the experimenter and subject were two people who took turns switching roles: the subject becoming the experimenter who subjected the former experimenter to test. This contrasts dramatically with the subsequent notions of an objective psychology, in which the subject is thought of as an object *s* that must be scrupulously set apart from the experimenter in order to avoid contamination.

Secondly, Danziger's book is about constructing a family of practices within which the subject is embedded. The upshot is a laboratory that is expanded to occupy the worlds of business, the military, education, law, and pathology, where people are regarded as subjects for testing. In a powerful passage at the end of his book, Danziger writes of "a fundamental convergence between contexts of investigation and contexts of application":

> the individuals under investigation became the objects for the exercise of a certain kind of social power. This was not a personal, let alone violent, kind of power, but the kind of impersonal power that Foucault has characterized as being based on "discipline." It is the kind of power that is involved in the management of persons through the subjection of individual action to an imposed analytic framework and cumulative measures of performance. The quantitative comparison and evaluation of these evoked individual performances then leads to an ordering of individuals under statistical norms. (Danziger 1990, 170)

A third item to be constructed is knowledge. (Danziger's last chapter is titled "The Social Construction of Psychological Knowledge.") The passage quoted above continues as follows:

> Such procedures are at the same time techniques for disciplining individuals and the basis of methods for producing a certain kind of knowledge. As disciplinary techniques the relevant practices had arisen during the historical transformation of certain social institutions, like schools, hospitals, military institutions, and, one may add, industrial and commercial institutions. . . . This kind of knowledge was essentially administratively useful knowledge required to rationalize techniques of social control in certain institutional contexts. Insofar as it had become devoted to the production of such knowledge, mid-twentieth-century psychology had been transformed into an administrative science.

Only by implication does Danziger discuss a fourth category, individual people. We are now trained to answer questionnaires or perform various tasks in order to find out our talents or what ails us. Of course the tests themselves do not settle things. Some readers will wish I had followed the advice given after my vocational aptitude tests early in high school—that I should become a meteorologist. The point is not what the tests say about each of us, but that each is now a kind of person who hardly existed a century and a half ago: fit subject for testing. Without us as common fodder for tests, there could hardly be such a thing as the *Mental Measurements Yearbook* (Mitchell, 1992). This handbook is scrupulous in admitting only very well validated and widely used tests. (Meehl's construct validity is strictly enforced!) The number of available tests has doubled with each edition over the past decades.

Danziger's book is a paragon of fairly literal constructionism. It presents a history of crafting various parts that are in turn assembled into larger structures. Experimental psychology begins with the physiology laboratory as model. Through the use of that model a new type of investigation is constructed. Certain types of inquiry are pared away from it—Wundt's introspection, for example. A new element is added. Subjects are not treated individually; aggregates become essential as statistical technologies are advanced. Statistical procedures from agronomy or biometrics are incorporated, often in black-box form; the psychologists who use the tests often have little idea of their rationale. There have been meta-experiments in which fully accredited psychologists were asked what a significance level means; only a minority give methodologically sound answers.

The metaphor of construction fits the chain of events that Danziger organizes. This is because there is something of a historical step-by-step building of specific techniques, institutions, and problems, each using previous steps, and assembled to form a further stage in the production of later techniques, institutions, and problems.

UNMASKING

Chapter 1 mentioned another metaphor, the metaphor of unmasking. It goes back to a familiar predecessor of constructionism—exactly contemporary with logical positivism. In his definitive 1925 paper on the sociology of knowledge, Karl Mannheim stated the four factors that created a need for the sociology of knowledge:

> (1) the self-relativization of thought and knowledge,
> (2) the appearance of a new form of relativization introduced by the "unmasking" turn of mind,
> (3) the emergence of a new system of reference, that of the social sphere, in respect of which thought could be conceived to be relative, and
> (4) the aspiration to make this relativization total, relating not only thought or idea, but a whole system of ideas, to an underlying social reality. (Mannheim 1925/1952, 144)

It is slightly misleading to take the term "unmasking" from Mannheim; for the word is that of his translator. The German original is *enthüllung,* which means revealing or exposing. In Wagner's *Parsifal* the cry goes up, "Uncover the grail!"—*Enhüllet den Gral!* "Unmasking" has, in addition, an overtone of exposing something that was deliberately covered, in order to conceal its true nature. The "unmasking turn of mind," wrote Mannheim, is

> a turn of mind which does not seek to refute, negate, or call in doubt certain ideas, but rather to *disintegrate* them, and that in such a way that the whole world outlook of a social stratum becomes disintegrated at the same time. We must pay attention, at this point, to the phenomenological distinction between "denying the truth" of an idea, and "determining the function" it exercises. In denying the truth of an idea, I

> still presuppose it as "thesis" and thus put myself upon the same theoretical (and nothing but theoretical) basis as the one on which the idea is constituted. In casting doubt upon the "idea," I still think within the same categorical pattern as the one in which it has its being. But when I do not even raise the question (or at least when I do not make this question the burden of my argument) whether what the idea asserts is true, but consider it merely in terms of the *extra-theoretical function* it serves, then, and only then, do I achieve an "unmasking" which in fact represents no theoretical refutation but the destruction of the practical effectiveness of these ideas. (Mannheim 1925/1952, 140)

Mannheim's model was Marxian, and he thought in terms of unmasking entire ideologies. He had, moreover, a sort of functionalism in mind. An ideology would be unmasked by showing the functions and interests that it served. Yet unmasking, in very much the terms used by Mannheim, has broader implications.

Mannheim wrote that the hidden history of the unmasking turn of mind "still calls for more exact investigation" (141). There is a lot of not-so-hidden history, featuring such household gods as Hegel, Marx, and Freud. An instructive hidden history would take in not only the unmasking of ideologies, but the local unmaskings attempted by Bertrand Russell and his admirers. The Russellian doctrine of logical constructions did not in general aim at refuting claims about theoretical or abstract entities, but instead tried to remove extra-theoretical presuppositions of statements about them.

Constructionism today is usually a more local sort of unmasking than Mannheim had in mind. Undoubtedly, studies of the construction of gender want to unmask an ideology. But let us turn to a more typical and less discussed example.

Serial Killers

Here is a set of common beliefs about serial killers. Serial murders are monstrous—far more crimes thus classified occur in the United States than elsewhere—the number of serial killers has been on the rise in many countries—serial killers are rare nonetheless—most but not all serial killers are men—these murderers had vile childhoods—their victims are chosen at random from a specific class of hapless people (prostitutes, black homosexuals, or whatever)—serial murder involves warped sex.

Every one of those beliefs is widely held. Each is, in the main, true. Together they form objective knowledge about a class of crimes, established by experts. Or so we think. Then we come across Philip Jenkins's *Using Murder: The Social Construction of the Serial Killer* (1994). We know what to expect. The author will not strictly refute our beliefs. But he will teach how the classification has been made up. He will show that the categorization of certain crimes as serial homicides functions for the benefit of some elements of law-and-order enforcement, and he will tell us how a new kind of expertise has come into being.

The effect of this is somewhat unsettling. It is not at all clear what to do, or that anything should be done. Take this true anecdote: a successful free-lance businesswoman told me that she will not let a courier with a package into her premises, especially when her attractive young assistant is about. There are too many serial killers out there. Her office is on the fourth floor of an upmarket mixed-use building in central, well-ordered Toronto. What is a relevant observation? At the level of truths about serial killers: they just don't invade premises like yours! Or at the unmasking level: you have somehow been conned into an irrational fear about a kind of person, a category constructed in order to serve certain interests, and to gratify certain fantasies! The anecdote is of no moment except as example. There may be straightforward political conclusions to draw from unmasking. Insofar as serial killing is an especially American conception (the British rippers and notorious Russian and Italian examples notwithstanding), is its "extra-theoretical force" intended to deflect attention from gun control, inner-city mayhem, and so forth?

What sorts of things are, in general, to be unmasked? Above all, the unpleasant—disaster (Fowlkes 1982). Even when we pass from specific kinds of people, such as serial killers, to quite general attributes of people, we are not surprised to find the construction of anger (Miller 1983), or both danger (McCormick 1995) and dangerousness (Webster et al. 1985). The construction of joy or tenderness would astonish us. But the all-too-good are doubtless in for it: we would be disheartened, but not shocked, by a construction analysis of Médecins Sans Frontières. When I first wrote the previous sentence I added "or Mother Teresa." Hardly had the ink dried than there appeared Christopher Hitchens's (1995) sardonic book about the saintly lady.

Hitchens did not exactly expose her—another reason that "unmasking" is to be preferred to the original German word, which can be translated as "exposing." Unmasking is different from exposing; they work at different levels. When the American evangelist Jim Bakker was shown to be sexually involved with acolytes and to be salting away a fortune, he was exposed, not unmasked. The difference between unmasking Teresa and exposing Bakker is analogous to Mannheim's distinction between challenging the extra-theoretical effectiveness of a doctrine, and simply refuting it, showing it to be false.

Refuting

Mannheim distinguished refuting from unmasking. Refuting a thesis works at the level of the thesis itself by showing it to be false. Unmasking undermines a thesis, by displaying its extra-theoretical function. The distinction is not all that sharp, for some analyses that chiefly aim at refuting or discrediting may gain added cogency by showing how what is to be refuted or discredited was constructed in the first place.

The construction metaphor is severely weakened by not distinguishing pure cases of unmasking from mixed cases of unmasking and refuting. Two remarkable books by Donald MacKenzie illustrate the difference. His *Inventing Accuracy,* subtitled *An Historical Sociology of Nuclear Missile Guidance* (1990) unmasks, but it also refutes the claim of any cold warrior (or of today's sons of cold warriors) to have "correctly" defined missile accuracy. The measured comparisons of "our" with "their" missiles were proposed in order to satisfy various political or technical agendas.

The point is not that missiles are not sufficiently accurate to be lethal. The point is that exceedingly delicate, competing, and incompatible measures of accuracy are defined to cater to two distinct interests. The paymasters and the public must be convinced that our missiles deliver excellent accuracy per dollar, but also that enemy missiles are so accurate that we need to build yet more missiles, or else introduce multiple entry missiles that leave a large enough footprint (the jargon of the trade) to cancel out inaccuracies. MacKenzie's historical sociology shows how experts and the lay public are taken in by the assertions of the weaponeers, engineers and policy makers alike. We walk away from MacKenzie's book knowing that in terms of the accuracy debates themselves, the standard measures of accuracy correspond not to some ideal measure of accuracy but to the interests of the parties involved. The measures are better or worse insofar as they serve the goals of maintaining or expanding arsenals.

Contrast MacKenzie's *Statistics in Britain: The Social Construction of Scientific Knowledge* (1981). This is a fascinating account of how statistical knowledge was produced in order to satisfy certain class interests of Victorian and Edwardian England. Eugenics became a dominant research interest of the later part of that period, and was featured, in a major way, in the contributions of such influential pioneers as Francis Galton and Karl Pearson. But we do not leave this book with the sense that regression, correlation coefficients, or the chi-squared test have been refuted. They may still be abused. We know from Herrnstein and Murray's *The Bell Curve* (1994) that the use of these tests, to pass from correlation to causal claims about race, is alive and well. Correlation and chi-squared tests nevertheless remain cornerstones of statistical inference, and MacKenzie did not even think of dislodging them. People who take issue with Herrnstein and Murray do not offer new statistical technology; they claim those authors drew incorrect inferences from the statistics.

MacKenzie's missile book described the social construction of missile accuracy *and* refuted measures of accuracy. His statistics book described the social construction of statistical methods, and left those methods intact. Mannheim would not have called that unmasking. If these two books are run together as two undifferentiated works of all-purpose constructionism, their distinct merits and contributions are altogether lost.

I write with some feeling here, because of my own book about multiple personality. In one chapter (Hacking 1995b, ch. 9)I explained how a certain continuum hypothesis about dissociative behavior was set in place. It has become dogma that a tendency to dissociate—whose extreme form is multiple personality—forms a continuum. I described how this dogma became established by questionable psychological testing and abuse of statistical tools. Yes, I showed how the continuum of dissociation was constructed before our very eyes, a micro-social construction of a supposed psychological fact if ever there was one. But I also aimed at demolishing the evidence and the techniques. I hope I refuted the claim to fact. Because of the current enthusiasm for social construction, I have to say, pedantically, that the chapter in question was not a piece of constructionism, even though I described the willful construction of an unwarranted "pseudo-fact" by a small but very influential social group of psychiatrists and psychologists.

HUMAN AFFAIRS

In Chapter 3 I turn to construction ideas about the natural sciences. There is a body of such work, and it has recently attracted hostile attention, but it is as nothing compared to work on human affairs. Politics, ideology, and power matter more than metaphysics to most advocates of construction analyses of social and cultural phenomena. Talk of construction tends to undermine the authority of knowledge and categorization. It challenges complacent assumptions about the inevitability of what we have found out or our present ways of doing things—not by refuting or proposing a better, but by "unmasking." One area of focus involves people: childhood, gender, youth homelessness, danger, deafness, disaster, illness, madness, lesbianism, literacy, authorship. Another is kinds of person: the woman refugee, the child viewer of television, the psychologist's subject. There is also behavior, such as serial homicide or white collar crime, and feelings, such as anger. We have vital statistics and postmodernism. We can focus on these diverse examples in different ways. For example, youth homelessness is a condition; the homeless youth, or the runaway, is a kind of person.

Should we distinguish this great variety of items from kinds of inanimate entities, such as the quark, or knowledge about a tripeptide? Why are people different? We get an intimation of the answer from the motivation of much constructionism. Constructionists are greatly concerned with questions of power and control. The point of unmasking is to liberate the oppressed, to show how categories of knowledge are used in power relationships.

It is widely taken for granted in constructionist studies that power is not simply exercised from above. Women refugees or deaf people participate in and assist in the power structure. One hope of unmasking is to enable the deaf or the women refugees to take some control over their own destiny, by coming to own the very categories that are applied to them. I used to call kinds of people, kinds of human action, and varieties of human behavior by the made-up designation, "human kinds." It is an important feature of human kinds that they have effects on the people classified, but also the classified people can take matters into their own hands. I called this phenomenon "the looping effect of human kinds" (Hacking 1995a). I now prefer to talk about interactive kinds.

The fundamental idea is almost too simple-minded. People are self-conscious. They are capable of self-knowledge. They are potential moral agents for whom autonomy has been, since the days of Rousseau and Kant, a central Western value. Quarks and tripeptides are not moral agents and there is no looping effect for quarks. Hence constructionism applied to the natural sciences was in the first instance metaphysical or epistemological—about pictures of reality or of reasoning. When applied to the moral sciences, the interest must first of all be moral. Assuredly there are infirm boundaries. The nonhuman may increasingly be invested with moral qualities—species, forests, ecosystems, Gaia. Yet the modeling of the moral remains firmly rooted in human values and the potential for self-awareness. Although many constructionists are moved by deeply moral concerns, all-purpose talk of social construction has tended to deflect attention from moral issues. This is doubtless partly because of a nervousness, noticed in some constructionists, in admitting the possibility of the very idea of morality. But if the point of the exercise is moral, one should not be squeamish about saying so.

THE NATURAL SCIENCES

Karl Mannheim had an attitude to physical science very different from that of modern constructionists. "Scientific-technical thought," he wrote, "completes just one and the same system during successive periods . . ."

> Because it is the same system that is being built up in science in the course of the centuries, the phenomenon of change of meaning does not occur in this sphere, and we can picture the process of thought as direct progress toward ultimately "correct" knowledge which can be formulated only in one fashion. In physics, there are not several different concepts of "force," and if different concepts do appear in the history of physics, one can classify them as mere preparatory steps before the discovery of the correct concept prescribed by the axiomatic pattern of the system. (Mannheim 1925/1952, 170)

This attitude is characteristic of sociology of knowledge from Durkheim through Mannheim. It took individuals trained in the sciences to apply sociology to the sciences themselves. One such was Ludwik Fleck, a remarkable epidemiologist and immunologist, who published over 100 medical research papers, some written in the Lvov ghetto until it was destroyed in 1943. He was a survivor. In 1935 he published a path-breaking book about scientific thought-styles *(Denkstile),* and about the origin and development of a scientific fact (Fleck 1935/1979, Cohen and Schnelle 1986) In retrospect he looks like the first author to have had a thoroughly "constructionist" attitude to scientific facts, although blessedly he did not use the construction metaphor. It would not have been very apt—or literal—for his story of the Wasserman test for syphilis.

It is part of Fleck's thesis that scientific facts exist only within styles of thinking, a doctrine to which I am myself sympathetic (Hacking 1992b). Fleck did not allude to Mannheim, but he did write caustically of sociologists such as Durkheim, Lévy-Bruhl, and less well-known figures such as Gumplowicz and Jerusalem: "All these thinkers trained in sociology and classics, however, no matter how productive their ideas, commit a characteristic error. They exhibit an excessive respect, bordering on pious reverence, for scientific facts" (Fleck 1979, 47). The era of excessive respect has passed! That is one reason for the science wars of today. Scientists feel inexorable laws of nature are not treated with sufficient respect by the sociologists. In fact, the early sociologists did treat laws with complete respect, and accepted the background scientific ideology without question. Only a real scientist such as Fleck could start questioning the mystique in which he himself had been educated.

Leaving the subject of pious reverence for later, let us try to catch a glimpse of where Fleck was directing us. Here is one rather conservative way to understand the thrust of his and subsequent arguments. The standard view is of science as discovery of facts that exist "in the world." The world comes structured into facts. That is not a scientific hypothesis. It is a metaphysical picture.

Fleck had a different metaphysical picture. He wrote of the emergence and development of scientific facts. He did not mean just that they emerge in human consciousness and develop in the history of science. He meant that the world does not come with a unique prepackaged structure. If we want an old name for this metaphysical picture, it is nominalism.

Constructionism about the natural sciences is also, in part, a metaphysical position. It is directed at certain pictures of reality, truth, discovery, and necessity. It joins hands very naturally with what Nelson

Goodman calls irrealism: not realism, not anti-realism, but an indifference to such questions, which in itself is a metaphysical stance. Since neither scientists nor constructionists dare to use the word metaphysics, it is not surprising that they talk past each other, since each is standing on metaphysical ground in opposition to the other.

Talk of metaphysics will seem, to many, a highbrow evasion of the issues current in the science wars. On the contrary, it is a central part of the story, and ignorance of it brings confusion. But it is only part of the story. Already, in 1935, Fleck was challenging pious reverence for the sciences. After 1945 there was a backlash against science itself. Science had been at the service of the concentration camps and gas chambers; only science could have created Hiroshima and Nagasaki. There were valiant attempts to defend the value of science as a human endeavor. The most notable was Jacob Bronowski's television series, *The Ascent of Man.* This was shown for uplift to millions upon millions of English-speaking schoolchildren. It began with heartfelt concern. Here I, Jacob Bronowski, am a man whose relatives perished in the camps. I have made a pilgrimage there. Here I, Jacob Bronowski, am a man who helped pioneer operations research (as the theory of efficient bombing) for the Royal Air Force during the war. I have made a pilgrimage to Hiroshima. But I now want to restore the Enlightenment vision of science, as one of the greatest endeavors of the human race, which shall save us yet, when undertaken with humility. Science can be restored to a state of grace.

There was another reaction, what Richard Bernstein names "the rage against reason." A rage against science and scientists. A rage that continued through the nuclear arms race, the Doomsday machine, chemical weapons, ecological disaster, the silent spring, the nuclear winter. That rage was so powerful that it needed few allies, but in intellectual and academic circles it latched on to the metaphysics of constructionism.

That is because metaphysics can have ideological consequences. The sciences, for some researchers, seem to involve getting to know the essence of creation, the mind of God. The metaphysics of constructionism denies that creation had an essence, or that there is a God's eye view. It is a threat to such a world view. Likewise, feminist critics of the natural sciences formed alliances with constructionists, in order to undermine the idea that the sciences must proceed along an inevitable, preordained patriarchal track.

Constructionism about the natural sciences is not necessarily political or critical. A constructionist could be committed to the current enterprises of the natural sciences, and just as full of admiration for past genius and present achievements as the most gung-ho science journalist who weekly announces the latest discovery. But constructionism can be used to unmask an ideology of science, an ideology that is intended to produce pious reverence. It must be said, as a purely anecdotal generalization, that every single constructionist about the natural sciences whom I know well is thoroughly irreverent.

The science wars, as I see them, combine irreverent metaphysics and the rage against reason, on one side, and scientific metaphysics, and an Enlightenment faith in reason, on the other. Hence the next chapter is about metaphysics and rage.

Chapter Three

WHAT ABOUT THE NATURAL SCIENCES?

Is there any point in talking about social constructs in connection with the natural sciences? Yes, there is a point in doing so, but that may not be the best way to examine the issues. We should separate out some fundamental disagreements about natural science that are made contemporary by using the phrase "social construct." Call them *sticking points.* They begin with philosophy and go almost as far as politics. Many would prefer to proceed the other way round. Dorothy Nelkin (1996) wrote a one-page essay asking "What are the science wars really about?" Her answer is that "current theories about science do seem to call in question the image of selfless scientific objectivity and to undermine scientific authority, at a time when scientists want to claim their lost innocence, to be perceived as pure, unsullied seekers after truth. That is what the science wars are about."[1] Or, more dramatically, the science wars are fueled by the rage against reason-masquerading-as-innocence. We should never forget that, but neither rage nor an image promises a clear view of constructionism about science. We first must grasp some basic philosophical issues that separate the two sides.

The issues may be irresoluble, for they are contemporary versions of problems that have vexed Western thinkers for millennia. I shall deliberately avoid traditional formulations, because old words tend to become ancient hulks encrusted with barnacles. But scrape off the parasites for yourself, and you might glimpse the gleaming hull of an Aristotle or a Plato shining through. My observation is not that we ought to be doing the same old things that they began, but that the same old things are still being done.

Only towards the end of this chapter will I get around to two less highbrow and more politically engaged confrontations. One is in the spirit of Nelkin's diagnosis, and comes from parties to constructionism who challenge a comfortable image of science. The other, in a spirit of symmetry, comes from the scientific side, and expresses betrayal.

What Are the Natural Sciences?

"Natural science" and "social construct" are the keywords. There is no need to define the natural sciences because the old favorites, chemistry and physics, and the new favorite, molecular biology, will do. They are the sites where the battle must be joined. We are not surprised to hear that the results of primatology bear strong traces of their discoverers. We can well imagine what Donna Haraway (1989) and others have taught us in detail: accounts of the behavior of primates reflect the societies of the scientists who study them. We all know the bad jokes about British apes with stiff upper lips, ruthlessly enterprising American apes, hierarchical and communitarian Japanese apes, promiscuous French apes. Primates, perhaps, have been a field for working out ourselves as much as describing animal communities. But many readers blanch when they come across the idea that the results of physics, chemistry, and molecular biology are social constructs.

Who Are the Social Constructionists about Science?

Trevor Pinch and Wiebe Bijker (1987, 18–19) call all recent work in Science and Technology Studies "social constructivist." I shall be more narrow and literal. My two exemplars of social construct thought have already been mentioned quite often.[2] Both have "construct-" in the title or subtitle: Pickering's (1984) *Constructing Quarks,* and Latour and Woolgar's (1979) *Laboratory Life: The Social Construction of a Scientific Fact.* Old books, for sure, but ones whose authors are vigorously at work, and who are almost universally regarded as constructionists. Scientists reported in each book got Nobel prizes, so this is first-class science; no shoddy goods on view here. It is a further convenience that the two books target the natural sciences just mentioned. One is about high-energy physics, the other about organic chemistry.

My examples share a feature that may prompt suspicion. Some knowledgeable scientists quite like the books. Have not the authors sold out? For example, the longest critical notice of *Constructing Quarks* says that no one has any excuse for not understanding the basics of the high-energy physics of the 1970s. The reviewers state that despite the fearful constructionist ideology to be found in a couple of chapters, Pickering's book is a first-rate history and explanation of the subject, accurate and readable at the same time (Gingras and Schweber 1986). Latour and Woolgar worked in the Salk laboratory founded by Jonas Salk of the polio vaccine. He wrote a preface for their book, bemused but admiring. He had no problem with Latour and Woolgar's description of activities in the laboratory he founded.

I like that; it is important that accounts of laboratory science, no matter how subversive their intent, should on the surface sound realistic to people who know the field in question. But does not the very fact that a physicist says Pickering's book is quite good history of physics, or the fact that the patron of the laboratory liked Latour's version of events, show that the authors are not critical enough? I think not. Both Latour and Pickering have been reviled by men on the other side in the science wars. For some thinkers, they are public enemies numbers one and two.

My choice of examples may be criticized on other grounds. There is an entire group of fields named Sociology of Scientific Knowledge, Science and Technology Studies, and Social Studies of Science. Practitioners are widely lumped together as "constructionists," despite the fact that construction, per se, does not loom large on their agendas. Should I not use them as my examples?

There is the Edinburgh school, including Barry Barnes (1977, 1995) and David Bloor (1976). It became famous early for its "Strong Programme in the Sociology of Knowledge."[3] Lewis Wolpert (1993), the distinguished British embryologist and public spokesman about science, connects the Strong Programme with social constructionism. "Those who hold to the Strong Programme believe that all knowledge is essentially a social construct, and so all science [good or bad] merits the same attention" (p. 110). I have not found this argument (the *A*, therefore *B*) in the writings of Barnes or Bloor. I shall mention their symmetry thesis later, but constructionism does not seem to be so intimately involved in the Strong Programme as is commonly made out. We come to the Strong Programme chiefly at sticking point #3, where we reflect on the stability of some scientific knowledge. The Edinburgh school wants to explain it by considerations which most scientists consider to be external to what is known, that is, to the content of the science.

Then there is the Bath school, including Harry Collins (1985, 1990,

1998), Trevor Pinch (1986), (Collins and Pinch 1982, 1993). I have heard Collins described as the "gate keeper" of Sociology of Scientific Knowledge. Many other individuals also practice science studies with a slightly iconoclastic bent. David Gooding (1990), Karin Knorr-Cetina (1981), Michael Lynch (1985, 1993), Simon Schaffer and Steven Shapin (1985) (Shapin 1994, 1996). Latour and his colleague Michel Callon are held to be engaged in a slightly different project, named "actor-network theory." Latour's original co-author, Steve Woolgar, has gone off in other directions (1988), and has concerned himself with questions about how the social study of science, being a science, has theses that refer to itself—"reflexivity."

The fairly recent state of play among these workers can be found in Pickering's (1992) collection of specially commissioned papers. Should I not give all these alleged "constructionists" equal time? By using Pickering and Latour as exemplars, will I not skew things? Doubtless, but I prefer to skew things towards two workers, Latour and Pickering, who (a) were there in the beginning with *Construct*-titled books about specific branches of science, (b) whose work proceeds apace, at this very moment, in innovative ways, and (c) whose descriptions of laboratory science were held to be faithful, if idiosyncratic, by some scientists who knew the fields well—even when the philosophical conclusions of the books looked bizarre to the very same scientists. Finally, (d) they are held by some scientists to be public enemies.

Distinctions

"*I take it for granted* that science is a historically situated and social activity and that it is to be understood in relation to the *contexts* in which it occurs." So writes Steven Shapin (1996) in the introduction to his book on the scientific revolution. The excessive emphases suggest he is worried. I am not worried. So I can say it without emphasis: I take it for granted that science is a social activity, to be understood in its contexts. But only after a distinction!

What distinction? According to the physicist Sheldon Glashow (1992, 28), "the assemblage of these [universal] truths is what we call physical science." Well, an assemblage of truths, or even of falsifiable hypotheses, is not a social activity. So in Glashow's perfectly legitimate sense of the word, science is not an activity of any sort whatsoever. On the other hand, if by science we mean scientific activity, then it is (trivially) social.

Even those scientists who work mostly on their own have to communicate their work.

This distinction, between an activity and an assemblage of truths, does not beg any questions about social construction. But it does point to what should be at issue. Recall the distinction between process and product. For sociologists the processes of science, the scientific activity, should be the main object of study. But for scientists the most controversial philosophical issues are about science, the product, the assemblage of truths.

We must mind our distinctions. Most people dislike distinctions.[4] You may find that my discussion smells of the study. Why not just roll with the punches and talk straight? No. It is bad to cave in to careless talk and enthusiastic bravura. In a book review in *Nature,* Harry Collins (1995) recalls Richard Dawkins's statement that no one is a social constructionist at 30,000 feet. Dawkins, continues Collins, has money in his pocket up there in the friendly skies. And money is socially constructed! So how can Dawkins reject social constructionism? This tomfoolery allows us to state two home truths.

First, against Collins, nobody doubts that things whose very existence requires social institutions and contracts are social products. Nobody doubts that many things dear to us, including money, are the product of our society and our history, and require social practices to stay in place. Collins has ample ground to feel that he and his colleagues are misunderstood, but he seems to direct his spleen at the wrong target.

Second, against Dawkins. Many social constructionists about the natural sciences appear to dislike the sciences. Nevertheless, constructionists do not maintain that the propositions received in the natural sciences are in general false. They no not believe that artifacts, such as airplanes, engineered in the light of scientific knowledge, usually fail to work. Constructionists are creatures of Humian habit. They expect airplanes to get you there, and know that science, technology, and enterprise are essential for air travel. Dawkins has plenty to get mad about, but he too seems to direct his spleen at the wrong target.

What is true is that many science-haters and know-nothings latch on to constructionism as vindicating their impotent hostility to the sciences. Constructionism provides a voice for that rage against reason. And many constructionists do appear to dislike the practice and content of the sciences. When Collins (1993, 262) insists that "Most of us love science, include Einstein among our top five all time heroes . . ." (and on and on in a sentence with 65 more words), one cringes and mutters something about protesting too much. But Pickering and Latour manifestly like the science they study and do not have to say so. They may query some self-serving images of science that are in circulation, and exalted pictures of what scientists do, why they do it, and how they do it. That is very different from doubting the truth or applicability of any propositions widely received in the natural sciences.[5] If they are social constructionists, they are so at 30,000 feet.

Sometimes making a distinction can put an end to controversy: the opponents were speaking of different things, and there is no real conflict. On other occasions distinctions can foster dissent. In Chapter 1, and below, in Chapter 5, I try to make sense of the claim that something can be both real and a social construction. That is a conciliatory gesture. In this chapter I pursue the opposite strategy, of finding irresoluble differences between realists and constructionists. This is because the science wars are founded upon, among other things of a more political or social nature, profound and ancient philosophical disputes. Thus my strategy here is the exact opposite of Sergio Sismondo. He is a peacemaker. One "reason for the lack of realist/constructivist debate lies in the fact that each side usually views the other position as obviously untenable" (Sismondo 1996, 10). By lopping off extremism on the edges of both doctrines, he hopes to find common ground. In contrast, my sticking points emphasize philosophical barriers, real issues on which clear and honorable thinkers may eternally disagree.

STICKING POINT #1: CONTINGENCY

The boldest title in the natural science arena is *Constructing Quarks.* Pickering plainly meant social construction. But according to the Standard Model, quarks are the building blocks of the universe! How then could they be constructed, let alone socially constructed?

When someone speaks of the social construction of *X*, you have to ask, *X* = what? A first move is to distinguish between objects, ideas, and the items named by elevator words such as "fact," "truth," and "reality." Quarks, in that crude terminology, are objects. But Pickering does not claim that quarks, the objects, are constructed. So the *idea* of quarks, rather than quarks, might be constructed.

That is a bit of a let-down. Everyone knows that ideas about quarks emerged in the course of a historical process. To say that Pickering was writing about the idea of quarks, rather than the objects quarks, deprives his startling title of its novelty. That will not do. Pickering intended more than a history of events in high-energy physics during the 1970s, more than a history of ideas. What is this more?

One radical notion, which prompts talk of construction, is that Pickering does not believe that the emergence of the quark idea was inevitable. You have to be careful here. Obviously the march of high-energy physics was not inevitable—the debacle of the Super-Conducting Super-Collider reminds us of that. Funding might have ended in 1946. Gell-Mann, the quark-namer and author of *The Quark and the Jaguar* (1994), might have become the world expert on jaguars. When Pickering says that the actual development of high-energy physics was highly contingent, he intends us to think of something like high-energy physics as a rich and triumphant international science that evolved after World War II and is regarded as a tremendous success—but this imagined fundamental and equally successful physics does not proceed in anything like a quarky way.

Pickering does state some options that he believes were open to high-energy physics in the early 1970s (they are ably summarized for the lay reader by Nelson 1994, 538–540). He distinguishes what he calls the new physics from the old prequark physics that transformed high-energy work during the 1970s. The changes were not only in theory but also in instrumentation. The bubble chamber, which had long been the tool of preference for producing tracks of particle decay, was partially superseded by new kinds of detectors. Pickering thinks that the "old physics" could well have carried on, and that it was not predetermined that its vision of the world, and its methods of interfering with and interpreting the world, would cease to bear fruit. He argues that the old physics was in an important sense incommensurable with the new physics—a sense that is more perhaps precise than in Thomas Kuhn's writing.

Let us, however, attend not to the details but take the general claim: alternative "successful" science is in general always possible. What does successful mean? The standards of success in a science are partly determined by the science itself. If the standards of successful science are to some extent internal to a science, what can be meant by an equally successful but nonquarky fundamental physics? Successful by what criterion? One content-neutral criterion is Imre Lakatos's (1970) idea of progressive and degenerating research programs. A research program (in Lakatos's sense: he is not talking about research programs in the ordinary sense used in talking, say, about grant proposals) is a series of theories. For Lakatos a program is empirically progressive if the successive theories make new predictions not covered by predecessors, while retaining most earlier corroborated predictions. It is conceptually progressive if its theories regularly produce new concepts with rich and simplifying structures. We could add "technologically progressive" to the list of virtues. A program is degenerating if it lacks these virtues and if, when confronted by difficulties, it produces new theories that merely skirt the problems, saying, "none of our business."

I am not offering Lakatos's methodology of scientific research programs as a correct philosophy of science. It is one standing proposal that says, in a way that at present seems to be fairly neutral, what a successful branch of science is. It enables us to explain the notion of "an equally successful physics that did not proceed in a quarky way." We mean a research program that does not incorporate anything equivalent to the standard model, but which is as progressive as contemporary high-energy physics. It might even carry cosmology and the origin of the universe along with it, but with a different world view emerging, and nothing like a quark in sight. Most scientists think this is absurd. So here we have one substantive sticking point.

Pickering never denies that there are quarks. He maintains only that physics did not have to take a quarky route. His type of claim is quite general. Physics did not need to take a route that involved Maxwell's Equations, the Second Law of Thermodynamics, or the present values of the velocity of light. Applied mathematics did not need to pass through quaternions (a mathematical example of Pickering 1995a), and geology could have shunned dolomite (my final example in Chapter 7). Most scientists find such assertions ridiculous.

This sticking point is not about the truth, or reality, or whatever, of dolomite or Maxwell's Equations. But does not Pickering have to get down to questions about truth sooner or later? It is a merit of his approach that he leads to a basis for serious disagreement in which we need not (yet) become ensnared by philosophy-laden words like "truth." The two words with the biggest role in Pickering's recent work are *resistance* and *accommodation*.

Resistance and Accommodation

When Pickering wrote about high-energy physics, he was well aware of its materiel, the spacious accelerators, the intricate detectors, the problems of getting the beam running right. His more recent book, *The Mangle of Practice* (1995a), is perhaps the most materialist contribution to social studies of science to date. He examines a complex dialectic of theory, experiment, and above all the machinery, instrumentation, computing equipment, and so forth, the substance of the science. The old motto used to be, "Science proposes, nature disposes." People put up conjectures, test them in experimental situations, and nature gets rid of the ones that are false. Pickering's view adds some much needed structure to that maxim. Research scientists have theoretical models, speculative conjectures couched in terms of those models; they also have views of a much more down-to-earth sort, about how apparatus works and what you can do with it; how it can be designed, modified, adapted. Finally, there is that apparatus itself, equipment and instrumentation, some bought off the shelf, some carefully crafted and some jerry-built as inquiry demands it. Typically, the apparatus does not behave as expected. The world *resists.* Scientists who do not simply quit have to *accommodate* themselves to that resistance. They can do it in numerous ways. Correct the major theory under investigation. Revise beliefs about how the apparatus works. Modify the apparatus itself. The end product is a robust fit between all these elements.

Robust Fit

Pickering's picture can be compared to a thesis advanced at the start of the twentieth century by the French physicist, philosopher, and historian of science, Pierre Duhem (1906/1954). Suppose that an experimental observation is inconsistent with a speculative conjecture expressed within the context of a theoretical model. That does not automatically refute the conjecture. For the observation is inconsistent only with the conjecture as it is used in the model, when taken together with auxiliary hypotheses about how the apparatus works. In the light of a negative experimental result, one is forced to revise, yes, but one can revise either the major theory under investigation or the auxiliary hypotheses about the apparatus. In Duhem's illustrative fable of an astronomer probing the heavens and not finding what is expected, the stargazer could revise the theory of the celestial vault *or* revise the theory of how the telescope works. Pickering adds: or rebuild the telescope.

When it comes to apparatus there is "the concrete instrument that [the scientist] manipulates," and a "schematic model of the same instrument, constructed with symbols by the aid of theories" (Duhem p. 155). In physics there is also what physicists call the phenomenology, the interpretation and analysis of experimental results; phenomenologists are responsible for the mesh between overarching physical theory and data. Duhem emphasized that we could change the schematic model. In modern physics, we can also revise the phenomenology. Pickering adds that it is also open to us to modify the concrete instrument—the telescope, or whatever.

The dialectic of resistance and accommodation sometimes comes to a temporary halt. Does this halt become a sort of permanent benchmark? Can it be used to manufacture reliable reproducible technology if wanted? If so, let us say that the fit between theory, phenomenology, schematic model, and apparatus is *robust.*

In ordinary English, this word means strong, or sturdy.[6] The idea is familiar. The fit between theory, phenomenology, schematic model, and apparatus is robust when attempts to replicate an experiment go pretty smoothly—and when other groups of workers, with new apparatus, new tacit knowledge, and a different experimental culture do not encounter important new resistance. I do not want to overemphasize the replication of experiments: more commonly, people try to improve on an experiment, not to repeat it (Radder 1995). I do not want to exaggerate the ease with which tacit knowledge is transferred (Collins 1985). I say only that there is an intelligible sense in which a fit between theory, phenomenology, schematic model, and apparatus becomes robust.

Contingency Means No Predetermination

To sum up Pickering's doctrine: there could have been a research program as successful ("progressive") as that of high-energy physics in the 1970s, but with different theories, phenomenology, schematic descriptions of apparatus, and apparatus, and a different, and progressive, series of robust fits between these ingredients. Moreover—and this is something badly in need of clarification—the "different" physics would not have been equivalent to present physics. Not logically incompatible with, just different.

The constructionist about (the idea of) quarks thus claims that the upshot of the process of accommodation and resistance is not fully predetermined. Laboratory work requires that we get a robust fit between apparatus, beliefs about the apparatus, interpretations and analyses of data, and theories. Before a robust fit has been achieved, it is not determined what that fit will be. Not determined by how the world is, not determined by technology now in existence, not determined by the social practices of scientists, not determined by interests or networks, not determined by genius, not determined by anything.

Contingency Does Not Mean Underdetermination

This vision must be sharply distinguished from Quine's famous notion of the underdetermination of theory by experience.[7] Quine observed that many incompatible theories are logically consistent with any given body of experience. Even if all possible data were in, there would still "in principle" be infinitely many theories that were formally consistent with such data. That is a logical point.

Pickering's point is not a logical one. He claims that, at any stage in research, it is not predetermined what will happen next. Even if it is predetermined that an experiment will not work as hoped, how it will not work, and more importantly how people will adapt to resistance, are not predetermined. What is to be done is not a matter of "choosing" a theory, but of meddling with theory, apparatus, and accounts of what the apparatus is doing. Pickering is talking about what will count next as data, what the research personnel will do, how the world will resist, what won't work, how the researchers will interpret that. None of that, in his view, is predetermined. Hence he is opposed even to the modest doctrine of Peter Galison that theoretical and instrumental traditions place constraints on the results of research (Galison, Pickering 1995c). The spat between Pickering and Galison has nothing to do with Quine's merely logical and hypothetical ideas. In his early work Pickering himself may have attempted some alliance with Quine in early work (1986, 5f, 404), but that was a mistake. His current analysis has nothing to do with Quinean underdetermination.

The constructionist believes that many robust fits were possible, although in the end only one seems conceivable. The actual fit that is arrived at is contingent. Physics did not have to develop in a quarky way. This is not because physicists, by some collective act of decision, could have wittingly chosen one account of the world over another. No such fanciful libertarianism is to be found in Pickering's work. The claim is that there are different ways of adapting to resistance, involving not only thinking, but also making different types of apparatus, and many ways of working in and adapting to the resultant material world.

The words "accommodate" and "adapt" immediately make one think of biological adaptation and evolution. One distinguished reviewer of Pickering's *Mangle of Practice,* John Ziman (1996), picked up the idea, and in a recent talk Pickering (1997) carried that forward. No set of conditions determines future biological evolution. In the same way, no set of conditions—including "how the world is"—predetermines the evolution of a science. In particular, in my terminology, they do not predetermine the shape of any robust fit that evolves.

Who might be troubled by contingency, so understood? Physicists, not metaphysicians.

Alien Science

Many physicists find it inconceivable, in retrospect, that there could have been a successful fundamental physics of the 1970s that did not take something like the quark road. Of course quarks are not the end. Perhaps there are lepto-quarks. Maybe quarks themselves drop out of the cosmotemporal mush to which our apparatus is directing us. But smart and well-supported groups addressing anything like the topics addressed by physicists in the sixties and seventies would inevitably have developed ideas very much like those that actually evolved. They reject Pickering's suggestion that the "old physics" and its detectors need not have been displaced. They agree that there is plenty of trivial contingency. Solemn names rather than joke names such as "quark" and "charm" might have been used, but the fundamental structure of any physics would be much the same. So would the material structure of apparatus, by and large. Some may even argue that the institutional structure would have to have evolved in something like the way it did, but most physicists are not interested in making claims like that.

Any successful science would have to have been equivalent to actual science. What does that mean? Some physicists take the transhuman stance, parodied by Donna Haraway (1991) as the God-trick. Here is Sheldon Glashow (1992, 28), co-winner of a Nobel prize in physics with Abdul Salam and Steven Weinberg: "Any intelligent alien anywhere would have come upon the same logical system as we have to explain the structure of protons and the nature of supernovae."

Glashow does not doubt, as part of his faith, that the alien would, if sufficiently intelligent, have hit on protons and supernovae as something whose structure needs explaining. Perhaps Pickering would query whether a successful alien physics would need to investigate protons, but I want to attend to a different problem. What is "the same" logical system? And what exactly does "logical" mean here? (We hope that Glashow is not using the word "logical" in a merely rhetorical way, without much more content than "jolly good.")

Glashow holds that any system of fundamental physics that emerged would in some important sense be equivalent to what we have arrived at (or will arrive at, after resolving remaining anomalies). But what sense is that? His fellow prize winner Steven Weinberg (1996, 14) offers an apparently operational test of equivalence. "If we ever discover intelligent creatures on some distant planet and translate their scientific works, we will find that we and they have discovered the same laws." Weinberg means, of course, the same laws of fundamental physics; those aliens might not even have the same biological make-up as we do, and hence not have hit on the same fundamental biological laws.

Philosophers have troubles with translation. There is Quine's doctrine of the indeterminacy of translation. A reader of Quine, or of Donald Davidson, might agree with Weinberg, but not to Weinberg's satisfaction. We find aliens speaking Alien. How do we know that Alien is a language at all? Only, says Davidson, if we can translate it, by and large, into our language. That requires (argues Davidson) that we assume that aliens share a good many beliefs with us. So we think we have translated the language of these beings only if we have translated their physics into something like ours. Hence translation begs the question of equivalence.

Or to use a thought that Quine used for basic formal logic, we would say that Alien sentences express statements of physics only if they are translatable into something recognizable as our physics. On that view of matters, Weinberg's claim turns out to be an empty tautology.

I have a lot of problems with this use of Quine or Davidson, but I do not see how to turn Weinberg's criterion into a substantive definition of equivalence. Weinberg (1996b, 56) has been more explicit. He says that Maxwell's Equations for electricity and magnetism must be deducible from any sound physics. Does deducibility do the trick?

There are several difficulties, one small, one large, and one curious.

First the small one.[8] World history could have been fundamentally different. Pascal, Leibniz, and above all Charles Babbage had the basic idea of the modern computer that has transformed the late twentieth century. Suppose (what is impossible) that Babbage had got it right in the early nineteenth century. Suppose we had something like massive high-speed Cray computers by 1850. Then the analytical mathematics in which Maxwell's Equations are cast would have been unnecessary. We could have bypassed Maxwell's Equations! On this fanciful hypothesis, it was not absolutely inevitable that physics took a Maxwellian route. Maxwell's equations would not even have been deducible.

Nevertheless, the physicist interjects, the formal structure of the computations done by the imagined Babbage Supercomputer would have in a certain sense conformed to what we call Maxwell's Equations—because that is how the world is. In my opinion, this notion of "conforming to" is even more obscure than the notion that any theory arising would be "equivalent to" Maxwell's, but let that stand.

The big difficulty with deducibility as a criterion of equivalence is at a different level. Figuring out the deductions does not leave everything the same. Weinberg conveys a picture akin to the schoolchild set an exercise in Euclidean geometry. If the child solves the problem, she writes Q. E. D. at the end of her proof. *Quod erat demonstrandum.* In a developing science the *"quod"* is usually not there before the proof. The great figures of what was once called rational mechanics, men like Laplace and Lagrange working around 1800, were in some sense obtaining consequences of Newton's laws of motion and gravitation. But they had to invent the mathematics that would do it. They had to invent the language in which the conclusions could be expressed. They had to articulate the theory. They were not just joining up the dots to complete a picture. They had to put in the dots. I am here only pointing to enormously difficult questions. Deducibility, translatability, and equivalence are not transparent ideas.

The curious difficulty is best stated by another physicist, Richard Feynman. He was discussing three distinct presentations of what we now call the law of gravitation:

> Mathematically each of the three different formulations, Newton's law, the local field theory and the minimum principle, gives exactly the same consequences. What do we do then? You will read in all the books that we cannot decide scientifically on one or the other. That is true.

> They are equivalent scientifically. It is impossible to make a decision, because there is no experimental way to distinguish between them if all the consequences are the same. But psychologically they are very different in two ways. First, philosophically you like them or do not like them; and training is the only way to beat that disease. Second, psychologically they are very different because they are completely unequivalent when you are trying to guess new laws (Feynman 1967, 53).[9]

An older case is the claimed equivalence of Newton's and Leibniz's formulations of the differential calculus; one may argue that it is not just a matter of arbitrary choice that we ended up with the Leibizian vision rather Newton's doctrine of fluxions. A more familiar and modern case is the equivalence of wave and matrix mechanics. Long ago, Norwood Russell Hanson (1961) drew attention to the ways in which formal proven equivalence may still allow of different uses, goals, and understanding of different "formulations." These matters do not strike most physicists as troubling, but they do deeply perplex the historian. One of the things that happens, in the evolution of a science, is that functionally nonequivalent systems become, are made, equivalent, and all traces of the former nonequivalence are obliterated.

Philosophers have been a little more cautious than some physicists in formulating what Bernard Williams calls "an absolute conception of the world." Williams's project has, however, mainly been to draw a contrast between scientific and moral reasoning. In the course of making that distinction he writes that "In a scientific inquiry there should ideally be convergence on an answer, where the best explanation of the convergence involves the idea that the answer represents how things are" (Williams 1985, 136).[10] We could explain "convergence on an answer" in three distinct ways: small-scale, big-scale, and unique-ultimate. These distinctions are not germane to the point Williams was making, for he wanted to separate ethics and science. In ethics, even if there were convergence on an answer or answers to fundamental moral dilemmas, the best explanation for that convergence would not be that the answer represents how things are. But since someone might invoke Williams's absolute conception for the science wars, we should make more clear what could be meant by convergence.

Small-scale convergence—of course! In the sciences we converge on answers all the time, whenever we obtain robust fits between theory, phenomenology, models of the apparatus, and apparatus. There is nothing ideal about that. It is a regular achievement. What about the best explanation of getting a robust fit? "That represents how things are"? That is not helpful if someone really wanted to know why theory, experiment, and apparatus fitted together, but such an answer, if someone were to give it, does not challenge contingency. Pickering claims that "how things are" does not uniquely predetermine which robust fits are achieved, from day to day. Williams gives no reason to disagree.

Big-scale convergence: Williams shows in context that he has in mind not little real-life answers to scientific questions, but something more in the grand scheme of things. By a big answer does he mean that science should converge on *an* answer, or that there is one and only one answer upon which we could converge? If he meant only *an* answer, then the notion of contingency is altogether consistent with Williams's absolute conception of the world.

Unique-ultimate: perhaps Williams wanted us not to think of *an* answer upon which inquiry should converge. Perhaps he meant that, ideally, there is only one answer upon which we could converge, if we were to converge. Glashow (1992: 28) expresses the idea more poetically. He holds that there are "eternal, objective, ahistorical, socially neutral, external and universal truths, and that the assemblage of these truths is what we call physical science." He did not exactly say that there is just one such assemblage, but we are pretty sure that is what he intended.

Formally speaking, the contingency thesis is entirely consistent with the ultimate one-and-only picture upon which inquiry in the physical sciences will converge. For there could be many roads to the one true ultimate theory, or none at all. If there were many roads, then the physics at each way station on each road would be different from the physics at way stations on every other road.[11] Once again, Williams's absolute conception of the world does not really cross the contingency thesis. This is hardly surprising, for Williams had a different motivation, namely to state a fundamental principle to distinguish science from ethics.[12]

The Sticking Point

The constructionist maintains a *contingency thesis*. In the case of physics, (a) physics (theoretical, experimental, material) could have developed in, for example, a nonquarky way, and, by the detailed standards that would have evolved with this alternative physics, could have been as successful as recent physics has been by *its* detailed standards. Moreover, (b) there is no sense in which this imagined alternative physics would be equivalent to present physics. The physicist denies that. Physicists are inclined to say, put up or shut up. Show us an alternative development. They ignore or reject Pickering's discussion of the continued viability of the old physics.

The sticking point need not be at quarks. But some things definitely are noncontingent, say the physicists, and their appearance in physics was inevitable if the science was to progress at all. When the physicist's sticking point is placed under severe challenge, there are several fall-back examples: Maxwell's Equations, the Second Law of Thermodynamics, the velocity of light. The contingency claim is that neither the law nor the equations nor the velocity (nor anything equivalent) are inevitable parts of any science as successful as present science.

In ordinary philosophical language, necessity is the contrary of contingency. But it would be confusing to call the physicists who oppose the contingency thesis "necessitarians." My physicist protagonists are *inevitabilists.* They do not think that the progress of physics was inevitable (we could have stayed with Zen). They do think that *if* successful physics took place, *then* it would inevitably have happened in something like our way.

Truly metaphysical issues do not yet arise. Strictly speaking, the contingency thesis is formally consistent with any metaphysics. Perhaps that is irrelevant: we do not want to speak strictly and formally in this connection. This is because metaphysics must arise from a certain sense of ourselves in the world. And a sturdy sense of reality—is that not metaphysical?—may find the contingency thesis altogether repugnant. We don't live in the kind of world in which the contingency thesis could be true! That is no empirical exclamation, derived by inference from experience. It is, if not a built-in sensibility, a sensibility that arises in a great many people in Western civilization who are attracted to scientific styles of reasoning. If that is what you mean by metaphysics, then metaphysics is appalled at the very thought of contingency. I shall turn to that kind of metaphysics at sticking point #2. I take it seriously.

When we turn to the metaphysics of the schools, the contingency thesis appears to be consistent with any standard metaphysics. (So much the worse for the standards and the schools, you may say.) For example, contingency is consistent with the scholastic debating point of the 1980s called "scientific realism." Many versions of that doctrine state that physics aims at the truth, and if it succeeds, it tells the truth. If the physics refers to some type of unobservable entity, then, if the physics is true, entities of that type exist. Many social students of science reject any version of scientific realism. So do many philosophers, such as Bas van Fraassen (1980). But the contingency thesis itself is perfectly consistent with such scientific realism, and indeed anti-realists, such as van Fraassen, might dislike the contingency thesis wholeheartedly. Pickering (1995a, 171) has become so mellow that he says he is agnostic about what he calls correspondence realism. He is right. Scientific realism simply does not matter to what he cares about, namely contingency.

STICKING POINT #2: NOMINALISM

High-level semantical words like "fact," "real," "true," and "knowledge" are tricky. Their definitions, being prone to vicious circles, embarrass the makers of dictionaries. These words work at a different level from that of words for ideas or words for objects. For brevity I have called them elevator words. They are used to say something about what we say about the world. Facts, truths, knowledge, and reality are not in the world like protozoa, or being in love. Philosophers keep on fussing with them. Theories of truth and theories of knowledge produce endless books. From the early nineteenth century until the 1930s, epistemology was king. More recently, theories of truth have ruled the roost. It would be feckless to address such mighty topics here, for one is not going to make any quick progress. My plan is to change the vocabulary slightly, so that we do not go on saying exactly the same unfocused things. But first, facts.

Facts

Latour and Woolgar's *Laboratory Life* was originally subtitled, *The Social Construction of Scientific Facts.* It centers on a discovery in endocrinology. Latour had, as an ethnographer, studied one of the two laboratories in which the work was done. Many scientists believe that this book, and Latour's later *Science in Action,* demean their work and treat serious activity as a matter of personal aggrandizement and network building. I shall discuss that reaction briefly towards the end of this chapter. Here let us think about facts.

First a warning. Although I use Latour to introduce a discussion about facts, that is hardly the core of his subsequent work. He has recently been very clear about the center of gravity of his kind of science studies. "Instead of ideas, thoughts, and scientific minds," he writes in the preface to the new French pocket-book edition of *Science in Action,* "one recovers practices, bodies, places, groups, instruments, objects, nodes, networks" (1996, 14). In *Laboratory Life* there was a great deal of emphasis on one type of entity: inscriptions. Indeed we were told that the main products of a laboratory are inscriptions—preprints, graphs, traces, photographs, published papers, and now e-mail. *Science in Action* has, happily, a much more material vision of science.

Latour and Woolgar briefly emphasized etymology. The word "fact" comes from the Latin *factum,* a noun derived from the past participle of *facere,* to do, or to make. Facts, they said, are made. Since made things exist, Latour and Woolgar (1986, 180) did "not wish to say that facts do not exist nor that there is no such thing as reality." Their point was "that 'out-there-ness' is the *consequence* of scientific work rather than its cause." And: " 'reality' cannot be used to explain why a statement becomes a fact."

Philosophical purists like myself feel uncomfortable about statements "becoming" facts. Statements state facts, and scientific facts do not come into being. If they are facts, expressed by tenseless sentences, then they are facts, timelessly, and do not "become." I doubt that ordinary people are so uptight about the timeless character of facts and truths as philosophers. In Chapter 7 I quote Humphry Davy (1812, 3), that master of so many scientific trades, who talks about how, after rigorous testing, a conjecture "becomes scientific truth."

Analytic philosophers do have the strongest inclination to say that facts discovered in the natural sciences are tenseless and timeless ("eternal" as Glashow put it). That's harmless, unless we grant a peculiar explanatory power to these abstractions. Latour and Woolgar were surely right. We should not *explain* why some people believe *p* by saying that *p* is true, or corresponds to a fact, or the facts.[13] For example: someone believes that the universe began with what for brevity we call a big bang. A host of reasons now support this belief. But after you have listed all the reasons, you should not add, as if it were an additional reason for believing in the big bang, "and it is true that the universe began with a big bang." Or, "and it is a fact." This observation has nothing peculiarly to do with social construction. It could equally well have been advanced by an old-fashioned philosopher of language. It is a remark about the grammar of the verb, "to explain."

We need to be careful with words here and not confuse the philosophical idea of "correspondence" with quite ordinary and unexceptionable ways of talking. Someone may come to believe a hypothesis because "it fits the facts." The ordinary word "fits" does not mean the abstruse "corresponds to." We mean that some puzzling facts need explanation, and such and such a hypothesis is palatable, nay plausible, just because it jibes with or even explains those puzzling facts. To continue the example: the big bang theory was widely accepted in 1973, when it was seen to fit the newly discovered facts about uniform background radiation in the universe. Indeed some people came to believe the theory just because it fit the newly discovered facts. That explains why they changed their minds. But we should not explain why some people believe *p*, by saying that they do so because *p* is true, or corresponds to a fact, or the facts. When stated so cautiously, this conclusion about truth and explanation is not challenging. Anyone antagonistic to both the letter and spirit of constructionism could still agree that the truth of a scientific proposition in no way *explains* why people maintain, hold, believe, or assent to that proposition.

Nominalism

So what's the problem? A very old one: a contemporary version of an ancient debate between two metaphysical pictures of the relationship between thought and the world. Sticking point #2 is *nominalism.* There is a big danger in using a philosophical label that has been tossed around ever since Columbus sighted land in the Caribbean. (A philosophical dictionary says that the name "nominalism" came into circulation in 1492). Those who know the word will already understand it in their own way, while those to whom the word is unfamiliar, or for whom it merely reeks of tired old philosophy, will not want even to hear the syllables pronounced. Nevertheless it is part of my argument that the present science wars, especially as they hook up with social construction, have strong resonances with traditional philosophical issues.

Nominalism is a fancy way of saying name-ism. The most extreme name-ist holds that there is nothing peculiar to the items picked out by a common name such as "Douglas fir," except that those items are called Douglas fir. And the same goes for all names whatsoever. (The Douglas fir is a species of tree in the rainforest of the Northwest coast of North America, not a true fir at all, but named in honor of a British governor of British Columbia named Douglas. When the wood is sawed up into planks and unloaded at an English port, the English, who are inveterate nominalists, then call it pine.)

An unpleasant metaphor has been much used, in recent times, in this connection. People quote Socrates out of context and speak of carving nature at the joints. The Douglas fir, they say, is one of the joints of nature, at least in coastal British Columbia. Nominalists deny that nature has joints to be carved. Their opponents contend that good names, good accounts of nature, carve nature herself at her joints.

Rather than rehearse some history of philosophy, I shall try for a contemporary version of old issues of nominalism, tailored for questions about the natural sciences. Allow me two slightly romantic-sounding formulae. I want to convey the spirit of the division.

One party hopes that the world may, of its own nature, be structured in the ways in which we describe it. Even if we have not got things right, it is at least possible that the world is so structured. The whole point of inquiry is to find out about the world. The facts are there, arranged as they are, no matter how we describe them. To think otherwise is not to respect the universe but to suffer from hubris, to exalt that pip-squeak, the human mind.

The other party says it has an even deeper respect for the world. The world is so autonomous, so much to itself, that it does not even have what we call structure in itself. We make our puny representations of this world, but all the structure of which we can conceive lies within our representations. They are subject to severe constraints, of course. We have expectations of our interactions with the material world, and when they are not fulfilled, we do not lie about it, to ourselves or anyone else. In the fairly public domain of science, the cunning of apparatus and the genius of theory serve to keep us fairly honest.

What to call these two sides? I am content to say that the second party is nominalist. What about the first party? "Realism" once named the opposite of nominalism, but the word now means a lot of things, even in technical philosophy. One philosopher, preoccupied by issues raised by Michael Dummett, tells me that nobody nowadays uses "realism" as the opposite of nominalism. So I will take a name that because of its ugliness noone else will use, and speak of *inherent-structurism.* I suppose that most scientists believe that the world comes with an inherent structure, which it is their task to discover.[14]

The Sticking Point

The nominalist hopes only to be true to experience and interaction. The scientific nominalist is the more self-demanding, having to be true to the way in which apparatus does not work, having to accommodate, constantly, to the resistance of the material world. Nominalists are far more radical than the philosophers called anti-realists, who are skeptical about or agnostic about the unobservable entities postulated by theoretical sciences. Nominalists are not concerned with observability. They are as cautious about the needles of a fir tree as they are about electrons, when it comes to the inherent structure of the world.

Every person will describe the roles of these two different metaphysical pictures in different ways. I have tried to give a fair rhetorical shake to both. Various people have said somewhere or other that everyone is born either an Aristotelian or a Platonist.[15] Here, then, is an old and irresoluble ghost lurking behind much of the current folderol about social construction. The schoolmen named it nominalism, but they did not invent that cast of mind.

STICKING POINT #3: EXPLANATIONS OF STABILITY

It is striking how often Maxwell's Equations and the Second Law of Thermodynamics appear in debate, as if they were the last bastions of besieged scientists. They are said to be as real as rocks (I take on rocks in Chapter 7). One reason that they are so effective in argument is that they nicely move up and down in the trio of objects, ideas, and elevator words. They are like *objects:* they are in the world, are they not? If anything is "in the world," says the scientist, it is the Second Law and Maxwell's Equations. But the Law and Equations are also truly profound *ideas:* at the previous turn-of-the-century celebrations, the great American philosopher of science, and founder of pragmatism, Charles Sanders Peirce, said that the Second Law was the crowning intellectual achievement of the nineteenth century. In his famous lecture, *Two Cultures,* C. P. Snow asserted that every humanist should know the Second Law as a minimum literacy requirement. And finally, are not the Law and the Equations *facts?* And of course they are *knowledge.* Finally, they are *real,* "as real as anything else we know" (Weinberg 1996a, 14). The Law and the Equations are wonderfully fitted for rhetoric.

There is a more ordinary, and more important fact about the Second Law or Maxwell's Equations: they are not going to go away. And yet they could, in two ways. One, the universe itself could change (but we would not be here to witness that impossible cataclysm, for the human body is too frail to survive). Or the Law and the Equations would go away if we found out that they are false. That would be some scientific revolution.

The early years of the twentieth century witnessed many profound changes in physics: the theories of relativity; the quantum theories. Philosophers picked up on these novelties. Karl Popper taught that the sciences are in a permanent dialectic of conjecture and refutation. The best theories are the falsifiable ones. Thomas Kuhn took that one step further. He argued that the sciences pass through stages of radical change, followed by some transient stability he called "normal science." He even wrote of the necessity of scientific revolutions.

Future historians of the history and philosophy of the sciences may suggest that Popper and Kuhn worked in unusual times. Events early in the twentieth century made them think that science is essentially unstable. From now on (it is already being said) future large-scale instability seems quite unlikely. We will witness radical developments at present unforeseen. But what we have may persist, modified and built upon. The old idea that the sciences are cumulative may reign once more. Between 1962 (when Kuhn published *Structure*) and the late 1980s, the problem for philosophers of science was to understand revolution. Now the problem is to understand stability.

Stability has to be stated with caution and humility. Scientists have not become infallible, nor do they pretend to be. But there is the sentiment that a lot of science is here to stay. This is elegantly expressed by Steven Weinberg (1996a, 14) writing about Maxwell's Equations. I shall divide his statement into two parts, to be labeled [A] and [B]. [A] is the uncontroversial data. [B] represents what we are supposed to learn from the data. [B] shows that our classification by sticking points is quite useful. [B], which seems to be one point, is in fact several. Two of the points are versions of sticking points #1 and #2, while [B] also directs us to a third sticking point.

> [A]. None of the laws of physics known today (with the possible exception of the general principles of quantum mechanics) are exactly and universally valid. Nevertheless many of them have settled down to a final form, valid in certain known circumstances. The equations of electricity and magnetism that are today known as Maxwell's equations are not the equations originally written down by Maxwell; they are equations that physicists settled on after subsequent work by other physicists, notably the English scientist [and engineer] Oliver Heaviside. They are understood today to be an approximation that is valid for a limited context . . . but in this form and in this limited context they have survived for a century and may be expected to survive indefinitely.

Thus far, no one should take issue with one word of this statement. But we are perilously close to a host of issues.

Culture and Science

Norton Wise (1996, 55), the historian of nineteenth-century physics, did not take exception to what [A] stated but to one of the messages [A] intended to convey, namely, that Maxwell's Equations have nothing to do with human culture. They are just facts that we run up against. Wise argued that culture and science are inseparable. The Equations came "from the work of some of the most deeply religious people who have ever contemplated a battery: Oersted, discoverer of electromagnetism and author of *The Soul in Nature;* Faraday, devout member of the Sandemanian sect, who discovered electromagnetic induction and articulated field theory; William Thomson (Lord Kelvin) and Maxwell . . ." Other scholars would emphasize the role of empire, of the laying of telegraph cables below the seas, and across Persia to India, all of which had high priority in the minds of Kelvin, Heaviside, and Maxwell himself.

Weinberg (1996b, 56) retorted that "Whatever cultural influences went into the discovery of Maxwell's Equations and other laws of nature have been refined away, like slag from ore." The British Empire and Sandemanianism are mere curiosities of bygone days, perhaps still casting their shadows in the worlds of politics and piety, but not in the natural sciences.

The same sort of debate arises for the Second Law of Thermodynamics. Chapter 2 quoted Max Perutz (1996, 69) saying that the law is "an inexorable law of nature based on the atomic constitution of matter," which states that "heat cannot be transferred from a cold to a warm body without performing work." Perutz is one of the handful of people who created molecular biology (Nobel prize shared with John Kendrew for ribonucleic acid, or RNA, 1962). One of his later achievements was the structure of hemoglobin. Hemoglobin with its structure, he would surely say, is not a social construct. It is a fact of life, life itself, our lives. The history of its discovery, the history of Bragg's X-ray crystallography and later events, is a social history of science, including Perutz's leaving Nazi Vienna for England, his post-war collaboration with Kendrew, a British wartime physicist looking for greener pastures, and so forth. But hemoglobin is not a product of that history; it was there even before the emergence of the human race. Put that way, it sounds as if Perutz has to be right!

Some constructionists retort, in connection with the Second Law of Thermodynamics, "we have done the history, you scientists have not." (A less modest man than Perutz might say, I and a few others are the history of the discovery of the structure of hemoglobin, what do you mean, I have not done it?) But that does not really begin the debate, for the scientist says, the history, construction-as-process, does not matter. Yes, thermodynamics takes its name from the thermodynamic engine—the old name for the steam engine. Thermodynamics is vested in that ingenious centerpiece of the industrial revolution and wage capitalism. But the content of the Second Law, what it now means, is independent of its history. The Second Law still uses the concept of "work," which betrays its industrial origins, but that has no consequences for any present use of the Second Law.

Norton Wise made valuable points in criticism of Weinberg, but on this issue the scientists seem to win the day. Maxwell's Equations and the Second Law bear none of their history about them. The only possible case to make against the scientist's firm sense of timelessness is about the form, rather than the content, of electromagnetic theory. The very set of questions that led us to the Second Law or the Equations were formed by certain directions set by religion, empire, and industry. Given the questions, the content of the Law and the Equations was developed and became free of its history. But, it might be protested, the "form" of this kind of knowledge was historically determined, with great consequences for what we have found out. Not that what we have found out is false, but that the entire set of possible questions and answers in terms of which we think was only one option. The very form of what we have found out is not so free of history as scientists imagine. That is an interesting but very obscure idea. I try to explore it in Chapter 6 below. No one has shown how it would apply to the Law or the Equations. Lacking such an argument, we have to regard those icons, in their present form, as independent of their history.

A Big Jump

Steven Weinberg's passage [A] (read strictly and literally) does not bear on sticking point #1. [A] does not entail that physics had to develop along modern Maxwellian lines if it was to be successful (inevitabilism). [A] does not bear on sticking point #2. It does not entail that Maxwell's Equations are part of the inherent structure of the world. [A] says that the Equations have become stable and can be expected to "survive." Physicists will continue to accept them, use them, believe them, take them as favorite paragons of scientific knowledge. So far nothing controversial. Let us see what comes next. The paragraph continues:

> [B]. That is the sort of law of physics that I think corresponds to something as real as anything else we know. On this point scientists like Sokal and myself are apparently in clear disagreement with some of those whom Sokal satirizes. The objective nature of scientific knowledge has been denied by Andrew Ross and Bruno Latour and (as I understand them) the influential philosopher Richard Rorty and the late Thomas Kuhn, but it is taken for granted by most natural scientists.

With [B] we are seamlessly moved up by the elevator, with words such as "real," and "objective," and "knowledge." Those words ("As real as anything else we know") did not occur in [A]. Weinberg (1996b, 56) emphasized this point in a reply to criticisms: "I tried in my article to put my finger on precisely what divides me and many other scientists from cultural and historical relativists by saying that the issue is not the belief in objective reality itself, but the belief in the reality of laws of nature."

"As real as anything else we know": such words also spring naturally from the mouths of mathematicians doing number theory. The theorems are as real as anything we know. That means, first, as irresistible, as "inexorable" (Perutz's word for the Second Law of Thermodynamics) as anything we know. If you are going to think about these things at all, you are going to get *here,* to Maxwell's Equations, and also to the fact there is no greatest prime number and, late in the day, to Fermat's last theorem. That is an inevitability thesis, sticking point #1. Weinberg confirms this reading: "One of the things about laws of nature like Maxwell's equations that convinces me of their objective reality is the absence of a multiplicity of valid laws governing the same phenomena . . ."[16] Less graciously, contingentists who imagine an alternative successful science should put up or shut up.

As real as anything else we know. People who have never *experienced* a mathematical proof (the feeling of, as Wittgenstein put it, "the hardness of the logical must") seldom grasp what Platonistic mathematicians are on about. The sheer inexorability of mathematical proof has persuaded provers that the numbers and their properties are as real as, or even more real than, anything else we know. A physicist may have a similar experience in connection with Maxwell's equations. It is not that we have a lot of evidence that the Equations hold in certain domains. Yes, we have that, but that is not what gives the overpowering conviction that this is how things are, indeed, how things have to be. Weinberg is giving vent to this conviction, one of the deepest that a reflective human being can ever experience. Where we get to from that is to an inherent-structure thesis. When Weinberg states that Maxwell's Equations are as real as anything he knows, he means, among other things, that they are part of the inherent structure of the world. That takes us back to sticking point #2.

Thus [A] is uncontroversial, but it leads Weinberg to [B], which turns out to involve two distinct stances, both of which we have encountered already, namely our first two sticking points, contingency and nominalism. But before examining a third sticking point, let us look at one of the named figures with whom Weinberg disagrees, namely Thomas Kuhn. If taken at his word, he would surely doubt [A]. For he wrote of the necessity of scientific revolutions. He thought that a science could not remain lively unless from time to time it was shaken up by revolution. This is a very different perspective from Weinberg's. Norton Wise drew attention to Weinberg's astounding statement that "as far as culture or philosophy is concerned the difference between Newton's and Einstein's theories of gravitation, or between classical and quantum mechanics is immaterial" (as if Kant were no figure in culture or philosophy). One can think of Kuhn and Weinberg looking down a spyglass in opposite directions. Kuhn magnifies tumultuous events in the sciences. Weinberg makes them minuscule in the grand scheme of things. But this is not the immediate point at which Weinberg takes issue with Kuhn.

In *Structure* Kuhn rejected the idea of scientific progress towards some one final vision of the world. What we see in the history of science is progress away from previous beliefs. Weinberg (1996b, 56) quotes from some of Kuhn's late writings, where Kuhn had said "it's hard to imagine . . . what the phrase 'closer to the truth' can mean." Kuhn (like Nelson Goodman, who calls himself an irrealist) went on to make plain that he did not think there is a reality which science fails to get at. The notion of reality is, on the contrary, idle. Weinberg disagrees. Here we seem to have moved back to sticking point #2. Kuhn was a nominalist, and Weinberg is an inherent-structurist.

I have just made an observation about Weinberg and Kuhn which is intended to respect both. Weinberg said he was trying to put his finger on differences between "cultural and historical relativists" on the one hand, and physicists like himself. He writes as if he is putting his finger on some ephemeral debate that has flourished these thirty years or so. I suggest his finger points at a pair of attitudes that have opposed each other for at least 2300 years. My "sticking-point" analysis is intended to emphasize that this is not the first time that deeply committed and honest persons have, well, stuck. There is also a further point at which they have stuck: the sources of stability.

External Explanations

Historians, philosophers, and sociologists of science have advanced all manner of explanations for the acceptance and persistence of a body of scientific belief and practice. Latour's work (singled out for mention by Weinberg in [B]) has emphasized the network of events and agents that lies behind an item of knowledge. If you doubt the item, you have to challenge endless other items with which it is linked, challenge an expanding host of authorities, undo a net of thousands of directly or indirectly cited experts and results. The Edinburgh school began by emphasizing the interests of scientific workers, which directed their research and molded their conclusions.

Here we move to questions of evidence and reason. Why is the Edinburgh school said to favor social construction? Because instead of reasons for belief, it offers social explanations for belief. If we took the metaphor of "construction" literally, we could hardly call the Edinburgh school constructionist, but they certainly emphasize the social. Latour, while saying more about how construction is done, de-emphasizes the word "social," saying we have never been modern, never in fact separated the social from the natural. To the uncommitted, all such writers emphasize factors in science which strike one as *external* to the content of the sciences they describe.

That is part of what Weinberg, in quotation [B], finds lacking in Latour. Does Latour deny "the objective nature of scientific knowledge"? Yes (for Weinberg), because Latour thinks that external factors are highly relevant to the stabilization of some beliefs as knowledge. Perhaps even the ultimate stabilization, the persistent survival of Maxwell's Equations. And it will do no good for a partisan of Latour to respond that of course he doesn't deny the objective nature of scientific knowledge. Latour has explicitly written even in the first book (Latour and Woolgar 1979) that he and his collaborators do not deny reality, facts, and (adds the partisan) "the reality of laws of nature." All such protests are in vain at the tribunal of the physicist, because Latour thinks that external factors are relevant to the stability of laws of nature, while Weinberg thinks they are irrelevant. That is the nub. That is sticking point #3: external explanations of scientific stability.

Rationalism and Empiricism

This sounds like a recent and ephemeral dust-up. But it is probably analogous to some versions of the opposition between empiricists and rationalists. Indeed, you can even cast historical debates between, for example, Locke and Leibniz in terms of external and internal. Leibniz thinks that the reasons underlying truths are internal to those truths, while Locke holds that (our confidence in) truths about the world is always external, never grounded in more than our experience.

I shall not push the analogy further. Alan Nelson (1994), like myself making heavy use of Latour and Pickering, wrote of what he called constructivists versus rationalists. Rationalists think that most science proceeds as it does in the light of the good reasons produced by research. Some bodies of knowledge become stable because of the wealth of good theoretical and experimental reasons that can be adduced for them. Constructivists think that reasons are not decisive for the course of science. Nelson concludes that this issue will never be decided. Rationalists, at least retrospectively, can always adduce reasons that satisfy *them*. Constructivists, with equal ingenuity, can always find to their own satisfaction an openness where the upshot of research is settled by something other than reason. Something external. That is one way of saying that we have found an irresoluble "sticking point." Nelson is right to use the word "rationalist" to name one side, drawing attention to a lineage. One mark of these two traditionally named attitudes is that the one favors internal understandings of what knowledge is, while the empiricist favors external explanations.

The Sticking Point

The constructionist holds that explanations for the stability of scientific belief involve, at least in part, elements that are external to the professed content of the science. These elements typically include social factors, interests, networks, or however they be described. Opponents hold that whatever be the context of discovery, the explanation of stability is internal to the science itself.

ANTI-AUTHORITY BY UNMASKING

My three sticking points are intellectual, philosophical, in the best senses of those words. But some other problematic points are less philosophical, and they play to the emotions more than the intellect. They are not irresolvable sticking points but, one might say, sticky points that provoke anger more often than debate.

For example, ever since Freud, at least, it has been a common piece of rhetoric to "diagnose" what really troubles your opponent. That makes for bad argument. Even if your opponents are positively ill, they may have reasons worth considering or positions worth acknowledging. Nietzsche went mad, but those who ignore what he wrote before his madness do so at their peril. When it comes to argument, I am loath to diagnose. But I will run through one common diagnosis of the science wars, not much removed from Dorothy Nelkin's assessment quoted at the start of this chapter. It is observed that famous American physicists lead one of the fronts, damning, among other things, the very notion of social construction. (In Great Britain it is not physicists so much as life scientists: Richard Dawkins, Max Perutz, and Lewis Wolpert.) Why would physicists be especially embattled?

It has long been common to distinguish two main branches of physics, high-energy physics and what used to be called solid-state physics, now called by the more abstruse name of condensed-matter physics. Most of the advances that have affected our daily lives are the product of solid-state physics, even at the level of quartz watches, liquid display crystals, and lasers that run my compact disc player and helped fix my eyes after I had gone blind. But ever since World War II, when one read in the newspaper or saw on television some striking story about physics, chances are it was high-energy physics.

For some fifty years high-energy physics was the queen of the sciences, fully funded thanks to supposed military applications that began with the atomic bomb. But with the end of the Cold War, the financing of high-energy physics was abruptly curtailed. For quite independent reasons, the new queen of the sciences became molecular biology. And suddenly solid-state people, ignored by the public for so long, are about to take the driver's seat, partly because of the richness of the applications of their fundamental research. New PhDs in high-energy physics cannot get jobs, and go to work on Wall Street (systems analysis at Goldman Sachs turns out to be not so different from work on very small particles). Even when the new solid-state PhDs cannot get academic research work, they are in demand by industry, especially by start-up companies where the risks are high, but where the profits may be immense.

That is where rhetoric enters. The high-energy physicists (it is argued) are unnerved by their sudden fall from favor. That is why they are kicking up a fuss about social construction and anti-scientism in general![17] In my opinion, even if this were true, it would leave untouched the important question of whether the fuss is well founded. It does, of course, help explain the timing of the fuss.

It is important to consider other factors besides the proportions of national treasuries spent on different kinds of basic research. Money helps, but self-esteem and the respect of others are far more important to living a life. High-energy physicists have to some extent lost their cultural authority. By this I mean not just the ability to command vast resources of money and talent, but also the conviction that their life work is deemed to be profoundly significant not only by their peers but also by their culture, or world culture at large.[18] Great poets mired in poverty may have cultural authority without patrons. Today molecular biology, biological medicine, brain science, and even computer science (despite talk of nerds) have far more cultural authority than physics.

These observations about cultural authority are important to sociologists of debate, but they leave untouched the issue of whether the high-energy physicists and cosmologists are right in their contentions against social constructionists. That is why I spent such a long time distinguishing three rather ancient and philosophical sticking points. But I cannot pretend that we should discuss only issues of high metaphysics.

In Chapter 1 I distinguished some six grades of constructionist commitment. Pickering is not an activist trying to abolish the idea of quarks and give us something better. He does not even want to reform the standard model, or gauge theory, except that, in the spirit of anyone trained in the field, he would like to improve it. Pickering, on my account so far, comes out as an ironist about quarks. Latour and Woolgar look like ironists about their tripeptide, Thyrotropin Releasing Hormone. So why are they taken to be critics of science, when at least Latour has gone out of his way to make a sardonic crack to the effect that, like every good Frenchman, he is filled with admiration for the achievements of science? Because there is a strong element of unmasking in the work of many constructionists.

Their target is not the truth of propositions received in the sciences, but an exalted image of what science is up to, or the authority claimed by scientists for the work that they do. I briefly explained Mannheim's idea of unmasking in Chapter 2. "The unmasking turn of mind" does not try to refute ideas, but to harm them by exhibiting their "extra-theoretical function." Constructionists believe that there is an extra-theoretical function for inevitabilism, inherent-structurism, and the rejection of external explanations of the stability of the sciences. These three serve an ideology of science, in something like the sense intended by Mannheim. They serve the world outlook of a certain social stratum, that of scientists who present themselves as the deepest probers of the universe, discoverers of ultimate truths.[19]

That social stratum does not include the broad mass of scientists, pure and applied, who tend to be a modest lot. Most scientists are fairly humble about their work, which they gladly admit is a string of tentative conjectures, temperamental apparatus, and nervous results. But when they, or the elder statesmen of science, look on the entire activity, a note of authority creeps in. Science has found out, by and large, how things are (we are told), how they must be, in the present state of things. There will be deeper accounts not yet discovered, but present science is, overall, as deep as it gets right now. Constructionists urge that this ideology has an extra-theoretical function: ensuring the cultural authority of science. The received wisdom is that scientists must not be challenged, because they are the deep probers of the inner constitution of things.

Thus what is to be unmasked is both a vision of underlying reality revealed by physics, and the associated claims to profundity of the entire endeavor. Here we have an acrimonious contretemps. Scientists feel deeply hurt, they feel that social constructionists do not take them seriously. It is no use social constructionists trying to cheer everyone up, saying that they love physics, but not for the wrong reasons. The wound has been inflicted.

This contretemps hangs together with the three sticking points. For example, the most knock-down defense of authority has always been metaphysics. The divine right of kings, taken more seriously than we can conceive of today, is a nifty way to ensure the authority of the sovereign. Constructionists want to unmask metaphysics as a bolster for the authority of the sciences. They also want to show that the present state of science was not the only inevitable upshot of dedicated inquiry into the material world that surrounds us. We achieve a robust fit between theories and apparatus, but the fit that we achieve is not the only one we might have arrived at. Contingency also undermines authority, not in the sense of casting doubt upon the propositions received in the sciences, but in the sense of challenging their claim to an unparalleled profundity. And finally, the survival of Maxwell's Equations is not to be explained only by factors internal to electromagnetism, quantum electrodynamics, and cosmology.

LEFT AND RIGHT POLITICS

A heartfelt ethical issue also arises. The traditional right/left spectrum of politics and alliances has run into problems. Although I did not find Sokal's spoof as interesting as most of its readers did, he did raise one genuine issue.[20] He lamented that he, as scientist, identifies himself as someone on the left, in support of the oppressed, while the mantle of the left has been stolen by people who write "theory," among whom he might count the authors of *Constructing Quarks* or *Laboratory Life*.

In terms of the unmasking of established order, constructionists are properly put on the left. Their political attitude is nevertheless very much not in harmony with those scientists who see themselves as allies of the oppressed, but also feel like the special guardians of the most important truths about the world, the true bastions of objectivity. The scientists insist that in the end, objectivity has been the last support of the weak. Here is a disagreement. It is a rather messy matter, a sticky point involving deep-seated but ill-expressed attitudes. Who is on the left?

I take this question very seriously, for I am deeply sympathetic to both sides. Some years ago, after a talk of mine about verisimilitude a freedom fighter of days gone by insisted on the extent to which objective truth is called for, as a virtue, when one is fighting tyranny. The enemy always tries to steal it (*Pravda* and *Trud* were once newspapers named after the noblest ideal, truth). The villains never could get away with that, so long as the last words are: "that simply is not true, liar!" My fighter would have hated those who want to unmask the values of truth, reality, and fact. They want, as he sees it, to remove the last ledge upon which freedom and justice can stand. I saw what he meant, and feel humble towards a man who really worked for the liberation of his people.

Nevertheless a serious issue is joined. Feminists feel most strongly that they well know about oppression. Left/right: what did that mean except an array of men in the French National Assembly! Forget it. They see objectivity and abstract truth as tools that have been used against them. They remind us of the old refrain: women are subjective, men are objective. They argue that those very values, and the word objectivity, are a gigantic confidence trick. If any kind of objectivity is to be preserved, some argue, it must be one that strives for a multitude of standpoints.

I have nothing to contribute to this debate, precisely because I am torn. Perhaps it is a generational thing. The values of that freedom fighter are part of my values, and they are values, in his generation, of one standpoint, in the end. But I also grasp the force of some of the critique, and am unable to synthesize my inclinations. I invite others to confess to these difficulties, and to refrain from dogmatism.

KUHN AND FEYERABEND

We cannot leave the sciences without mentioning these two eminences. Most people would guess that the flamboyant anarchist, Paul Feyerabend, was more of a constructionist than that somber revolutionary, Thomas Kuhn. I find the opposite. We now have a check list to see how constructionist each author is. #1 Contingency; #2 Nominalism; #3 Stability. Let us use it.

Kuhn did not mention social construction in his 1962 masterpiece, *The Structure of Scientific Revolutions.* The words were not common parlance until after Berger and Luckman's *The Social Construction of Reality* appeared in 1966. In one chapter he argued that progress in science is "away from" past science, rather than "toward" a right account of an aspect of the world. That is an exceptionally strong contingency thesis. Every revolution is contingent. Nothing determined that one ought to go the way one did. Normal science, in contrast, proceeds in a rather inevitable way. Certain problems are set up, certain ways for solving them are established. What works is determined by the way the world collaborates or resists. A few anomalies are bound to persist, eventually throwing a science into crisis, followed by a new revolution.

Kuhn's normal science follows its ordained route. Given the way the world is and the questions posed by normal science, and the achievement (the paradigm) on which a normal science models itself, the upshot of inquiry is rather inevitable. We are going to get the anomalies that will lead us to a new sense of crisis. (Pickering is far more radical than Kuhn! His account of what I call the contingency of robust fit is all in the realm of normal science.) But the aftermath of revolution, the new paradigm that shines ahead, is entirely contingent.[21] Nothing determines the upshot of crisis. That radical contingency generated the storm that greeted Kuhn's book in 1962.

Kuhn was also a nominalist. This has not excited much interest even among philosophers of the sciences; following Kuhn's publication, they got caught up in drab little case histories and debated questions about when a change in theory is rational. Kuhn himself definitely checks off at sticking point #2 as a nominalist.

Kuhn already entered into the discussion of sticking point #3. One suspects that if he did admit any stability in science, he would explain it at least partly on external grounds. And of course he would be dubious about any permanent stability in any active science.

Kuhn did a great deal to undermine the ideology of science. He did not deliberately write as an unmasker, trying to expose false authority. Kuhn the person was quite well disposed to authority. But Kuhn the book had the effect of unmasking the authority of science in a quite remarkable way. On the one hand we got normal science as puzzle-solving. Most of the time scientists do not probe the deep structure of the universe. They engage in a superior sort of crossword-puzzle activity. What a put-down! The moments of glory, on the other hand, the pinnacles of revolution from which new worlds could be seen, were not predetermined by reason or wisdom, and their triumph was ensured chiefly by the death of old scientists. That is a parody of Kuhn, of course, but not a malicious one. Lakatos's wicked taunt, "mob psychology," captures the way that many people read the book. The authority of science was unmasked as never before.

The great professed anti-authoritarian figure was not Kuhn but Paul Feyerabend. He did not, however, try to disintegrate the ideology of science by unmasking it. He simply opposed it. And he did not do so on anything recognizable as social constructionist grounds. He was far more direct than that. And in the preface to the third edition of *Against Method,* he explicitly deplores the ways in which the sociologists of science want to demystify science.[22] Thus Feyerabend was anti-authoritarian but not by Mannheimian unmasking.

Did Feyerabend subscribe to a contingency thesis? He did think it is far more a matter of choice than we imagine, what kinds of questions we ask, and whether we want to pursue scientific enterprises at all. He mocked the scientific establishment as more dogmatic and exclusive than was the Roman Catholic Church confronting Galileo. But he did not claim that people in pursuit of certain ends could, in their interactions with the world, have gone more than one way. If there is contingency it is at the level of the methodologies that are favored at one time or another. These are not predetermined, but once the methods are in place, then science carries on towards its destinations, or so he may have implied. Feyerabend was a wonderful pluralist. But pluralism does not imply contingency. This is because every route that human beings may choose may develop rather inevitably. He encouraged homeopathy, acupuncture, psychic research, and much else. Those are remarkably stable enterprises, and one could plausibly, if surprisingly, hold that each has evolved rather inevitably—given the places of human beings in the world. I see little reason to attribute a strong contingency thesis to Feyerabend.

What about the stability of science, sticking point #3? He did think that lots of scientists were stuck in dull routines, but he was just being a good Popperian when he said so. The only sticking point at which Feyerabend definitely checks off is his nominalism, apparent, say, in the appendix on archaic Greece in the first edition of *Against Method.*

In conclusion, Kuhn was certainly a nominalist, and Feyerabend was a nominalist by inclination. Feyerabend was anti-authoritarian, but not for reasons of social construction. Kuhn's book did unmask science, while Feyerabend challenged its authority at its own level, the very opposite of unmasking. On the key issue of social construction, namely contingency, it is Kuhn whom, without anachronism, we can call social-constructionist. And although in the early days of social construction talk Feyerabend would have egged it on, he never advocated the contingency thesis. By the time that social construction had become an orthodoxy of that branch of academic studies called "theory" (not theory about something, just theory) he would have jeered at it.

CHECK LIST

The three sticking points form a check list. Where do you stand on social construction? Score yourself from 1 to 5 where 5 means you strongly stick on the constructionist side, and 1 the opposite. For example, on my reading of Kuhn, he scores 5, 5, 5. Here are my own scores, as debilitatingly ambivalent as you may have come to expect. Do your own.

#1 Contingency: 2.
#2 Nominalism: 4
#3 External explanations of stability: 3[23]

Chapter Four

MADNESS: BIOLOGICAL OR CONSTRUCTED?

I now turn to a quite different field of conflict, also couched in terms of construction. Mental illness provides the most pressing example.

It is easy to be skeptical about many of the entries in contemporary diagnostic manuals. How about Intermittent Explosive Disorder? Certainly, some people fly off the handle all too easily, but do they suffer from a mental illness, IED? Or is this just some construct concocted by psychiatrists? We suspect that IED has to do with medicalizing disagreeable patterns of behavior. It is easily argued that IED is not a diagnosis but a disciplinary device. If someone said that Intermittent Explosive Disorder is a social construct, I might wince at the overuse of social-construct talk, but would understand roughly what was meant.

Other mental illnesses are what I call transient. I do not mean that they last only for a time in the life of an individual. I mean that they show up only at some times and some places, for reasons which we can only suppose are connected with the culture of those times and places. The classic example is hysteria in late-nineteenth-century France. There is multiple personality in recent America. There is anorexia—of which young women can die—which is quite local in its history; at present it is more virulent in Argentina than anywhere else. It is all too tempting to call these social constructs.

Here I will not discuss transient mental illnesses, which I examine, in a very different way, in my book *Mad Travelers* (Hacking 1998a), nor will I discuss the disciplinary diagnoses such as Intermittent Explosive Disorder. Instead, I will examine illnesses such as schizophrenia and conditions such as mental retardation. These, it will be said, contrast strongly with anorexia, in that they have been with the human race in most places and times. There are no mental retardation epidemics in Argentina, even if the various words used to describe the condition, such as "feeble minded," were used only at a specific time and place and very strongly reflect social attitudes and institutional practices. The name "schizophrenia" was invented only in 1908. So what? These are "real" illnesses or conditions. And yet, and yet, there is a minority that will say that these disorders—and not just our ideas about them—are social constructs. Very often arguments are expressly put as: *X* is real—No, *X* is constructed.

It is not only the "constructed" that confuses us here, but also the "real." Hilary Putnam (1994, 452) hit the nail on the head, when he wrote about a "common philosophical error of supposing that 'reality' must refer to a single super thing, instead of looking at the ways in which we endlessly renegotiate—and are *forced* to renegotiate—our notion of reality as our language and our life develops." One of the reasons that we become confused in debates about whether an illness is real or not is that we fail to attend carefully enough to the grammar of the word itself (cf. J. L. Austin 1962, 72; Hacking 1983, 47; 1995, 11). But in the special context of mental illness we have, for the past two centuries, been constantly renegotiating our notion of reality.

"Social construct" and "real" do seem terribly at odds with each other. Part of the tension between the "real" and the "constructed" results from interaction between the two, between, say, child abuse, which is real enough, and the idea of child abuse, which is "constructed." But that is not all. We can also confuse more complex types of interactions, which make some people think of antique dualisms between mind and body. These come out most clearly when we turn to the very habitus of mind and body, psychopathology. Most present-day research scientists take schizophrenia to be at bottom a biochemical or neurological or genetic disorder (perhaps all three). A minority of critics think that in important ways the disease has been socially constructed. I do not want to take sides, but to create a space in which both ideas can be developed without too much immediate confrontation—and without much social-construction talk either.

What difficult terrain we enter! One of the reasons that I dislike talk of social construction is that it is like a miasma, a curling mist within which hover will-o'-the-wisps luring us to destruction. Yet such talk will no more go away than will our penchant for talking about reality. There are deep-seated needs for both ideas. Nothing I could say would discourage anyone from talking either about reality or social construction. Hilary Putnam, just quoted, said something very useful, but it is not going to change the way that even those who read him talk about reality. So instead I shall suggest some other ways to think about questions posed by the ideas of social construction—and reality. There are many difficult questions to address, so it is good to start with something relatively easy to follow.

CHILDREN

The distinction between objects and ideas is vague but should by now be easy enough, even if there are many subdistinctions of some subtlety. Yet this basic distinction conceals a very difficult issue. The trouble is that ideas often *interact* with states, conditions, behavior, actions, and individuals. Recall Philippe Ariès's well-known *Centuries of Childhood.* In the wake of that book, childhood has been called a social construct. Some people mean that the idea of childhood (and all that implies) has been constructed. Others mean that a certain state of a person, or even a period in the life of a human being, an actual span of time, has been constructed. Some thinkers may even mean that children, as they exist today, are constructed. States, conditions, stages of development, and children themselves are worldly objects, not ideas.

Thus it may be contended that children now—take two small individuals named Sam and Charlie-boy—are different from children at some other time, because the idea of childhood—the matrix of childhood—is different now. It may be argued that the state in which Sam and Charlie find themselves now is different from what it was for their ancestors, or even their mothers, Jane and Rachel, when they were very young. Conversely, the idea of childhood may have changed from what it was long ago, if children now are different from children then.

To use a less grand example than the whole of childhood, there has been a historical succession of ideas: fidgety, hyperactive, attention deficit, and Attention Deficit Hyperactivity Disorder.[1] Perhaps the children to which these terms have been applied over the course of this century are themselves different. Perhaps children diagnosed with ADHD are different from the children once called fidgety—in part because of the theories held about them, and the remedies that have been put in place around their bad habits. Conversely, it may be that the resulting changes in the children have contributed to the evolution of ideas about problem children. That is an example of interaction.

I want to focus not on the children but on the classification, those *kinds* of children, fidgety, hyperactive, attention-deficient. They are *interactive kinds.* I do not mean that hyperactive children, the individuals, are "interactive." Obviously hyperactive children, like any other children, interact with innumerable people and things in innumerable ways. "Interactive" is a new concept that applies not to people but to classifications, to kinds, to the kinds that can influence what is classified. And because kinds can interact with what is classified, the classification itself may be modified or replaced.

I do not necessarily mean that hyperactive children, as individuals, on their own, become aware of how they are classified, and thus react to the classification. Of course they may, but the interaction occurs in the larger matrix of institutions and practices surrounding this classification. There was a time when children described as hyperactive were placed in "stim-free" classrooms: classrooms in which stimuli were minimized, so that the children would have no occasion for excess activity. Desks were far apart. The walls had no decoration. The windows were curtained. The teacher wore a plain black dress with no ornaments. The walls were designed for minimum noise reflection. The classification *hyperactive* did not interact with the children simply because the individual children heard the word and changed accordingly. It interacted with those who were so described in institutions and practices that were predicated upon classifying children as hyperactive.

INTERACTIVE KINDS

There is a big difference between quarks and children. Children are conscious, self-conscious, very aware of their social environment, less articulate than many adults, perhaps, but, in a word, aware. People, including children, are agents, they act, as the philosophers say, under descriptions. The courses of action they choose, and indeed their ways of being, are by no means independent of the available descriptions under which they may act. Likewise, we experience ourselves in the world as being persons of various kinds. To repeat a quotation from the start of Chapter 1, it is said that "the experiences of being female or of having a disability are socially constructed" (Asche and Fine 1988, 5–6). That means, in part, that we are affected by the ways in which being female or having a disability are conceived, described, ordained by ourselves and the network of milieus in which we live.

Here I am concerned with kinds of people, their behavior, and their experiences involving action, awareness, agency, and self-awareness. The awareness may be personal, but more commonly is an awareness shared and developed within a group of people, embedded in practices and institutions to which they are assigned in virtue of the way in which they are classified.

We are especially concerned with classifications that, when known by people or by those around them, and put to work in institutions, change the ways in which individuals experience themselves—and may even lead people to evolve their feelings and behavior in part because they are so classified. Such kinds (of people and their behavior) are interactive kinds. This ugly phrase has the merit of recalling actors, agency and action. The *inter* may suggest the way in which the classification and the individual classified may interact, the way in which the actors may become self-aware as being of a kind, if only because of being treated or institutionalized as of that kind, and so experiencing themselves in that way.

INDIFFERENT KINDS

The word "kind" was first used as a free-standing noun in the philosophy of the sciences by William Whewell and John Stuart Mill, some 160 years ago. Here I use it to draw attention to the principle of classification, the kind itself, which interacts with those classified. And vice-versa, of course, it is people who interact with the classification.

There can be strong interactions. What was known about people of a kind may become false because people of that kind have changed in virtue of how they have been classified, what they believe about themselves, or because of how they have been treated as so classified. There is a looping effect.

I have not defined "interactive kind," but only pointed. Kinds that are the subject of intense scientific scrutiny are of special interest. There is a constant drive in the social and psychological sciences to emulate the natural sciences, and to produce true natural kinds of people. This is evidently true for basic research on pathologies such as schizophrenia and autism, but it is also, at present, equally true for some but only some investigators who study homosexuality (the search for the homosexual gene) or violent crime (is that an innate and heritable propensity?). There is a picture of an object to be searched out, the right kind, the kind that is true to nature, a fixed target if only we can get there. But perhaps it is a moving target, just because of the looping effect of human kinds? That is, new knowledge about "the criminal" or "the homosexual" becomes known to the people classified, changes the way these individuals behave, and loops back to force changes in the classifications and knowledge about them.

The notion of an interactive kind is fuzzy but not useless. Plenty of classifications differ fundamentally from any of the human kinds just mentioned. Quarks are not aware. A few of them may be affected by what people do to them in accelerators. Our knowledge about quarks affects quarks, but not because they become aware of what we know, and act accordingly. What name shall we give to classifications like that? Too much philosophy has been built into the epithet "natural kind." All I want is a contrast to interactive kinds. *Indifferent* will do. The classification "quark" is indifferent in the sense that calling a quark a quark makes no difference to the quark.

Indifferent does not imply passive. The classification *plutonium* is indifferent, but plutonium is singularly nonpassive. It kills. It exists only because human beings have created it. (That is not quite true: it was once thought that transuranic evolution had never, in nature, got to plutonium; in fact natural plutonium has been identified). Plutonium has a quite extraordinary relationship with people. They made it, and it kills them. But plutonium does not interact with the idea of plutonium, in virtue of being aware that it is called plutonium, or experiencing existence in plutonium institutions like reactors, bombs, and storage tanks. So I call it indifferent.

Microbes, not individually but as a class, may well interact with the way in which we intervene in the life of microbes. We try to kill bad microbes with penicillin derivatives. We cultivate good ones such as the acidophilus and bifidus we grow to make yogurt. In evolutionary terms, it is very good for these benevolent organisms that we like yogurt, and cultivate them. But some of the malevolent ones do pretty well too. Disease microbes that we try to kill may as a class, a species, respond to our murderous onslaught. They mutate. There is some evidence for what is called directed mutation. Under environmental stress, such as lack of edible food (lactates) that can be ingested by the microbes in a culture, the microbes mutate in a nonrandom, species-beneficial way so that they can feed. Maybe that is how disease microbes so quickly become resistant to our poisons.

Long ago Mary Douglas (1986, 100-102) saw the drift of where I was going when she read a draft of my essay (1986).[2] Do not microbes adapt themselves to us, quickly evolving strains that resist our antibacterial medications? Is there not a looping effect between the microbe and our knowledge? My simple-minded reply is that microbes do not do all these things because, either individually or collectively, they are aware of what we are doing to them. The classification *microbe* is indifferent, not interactive, although we are certainly not indifferent to microbes, and they do interact with us. But not because they know what they are doing.

NATURAL KINDS

When philosophers talk about natural kinds, they take the indifference—in my technical sense—of natural kinds for granted. That is to be expected. At the end of this chapter I shall make heavy use of the natural-kind philosophy and semantics that we owe to Hilary Putnam and Saul Kripke. Their innovative ideas were, in one respect, very conservative. They are part of a tradition that reaches back into the industrial revolution, when William Whewell and John Stuart Mill put the idea of natural kinds into philosophical circulation. At that time, more than ever before in human history, the distinctions between humans and nature, minds and matter, took a distinctive turn. The earth became covered by active machines, made by and tended by people, but running more or less on their own, thoroughly active and somewhat autonomous. In the seventeenth century, mechanical watches and an automatic clock on the spire at Strasbourg had moved philosophers to flights of fancy. But those devices *did* nothing. They were semantic; they were signifiers; they told us the time. In the early nineteenth century, the steam engine at the pit head, the steam locomotive, the spinning jenny accomplished unimaginable feats.[3]

Nature, despite the way that Romantics were fascinated by the sublime and the wild, continued to be thought of as passive, at least in the laboratory. Nature was acted on by us, and now by our creatures, the machines. Hence the concept of the natural kind came into currency, as something indifferent. The things classified by the natural-kind terms favored in philosophical writing are not aware of how they are classified, and do not interact with their classifications. The canonical examples have been: water, sulphur, horse, tiger, lemon, multiple sclerosis, heat and the color yellow. What an indifferent bunch! None is aware that it is so classified. Of course people and horses interact. Black Beauty and Flicka were (fictional) horses that attended to the humans who loved them, who in turn attended to the horses they loved. In denying that *horse* is an interactive kind, I am not denying that people and horses interact. I am saying that horses are no different for being classified as horses. Indeed it will make a difference, in law and to a Shetland pony, whether ponies are classified as horses: but not because the ponies know the law.

Continua

More iconoclastic thinkers than the philosophers of natural kinds will argue that there is a continuum between indifferent kinds and interactive kinds. I sympathize, but suspect that there is less of a continuum than a lot of fuzzy edges. Or perhaps a number of different continua. What about Bruno Latour's (1993) conception of *actants,* or Kathryn Addelson's (forthcoming) notion of *participants?* What about Andrew Pickering's (1995a) attribution of agency to the material world of the laboratory and beyond, a world that resists us and to which we adapt? These ways of thinking will become increasingly popular in the future.

What about cyborgs? When the word "Cyborg" was first introduced (with a capital letter *C*) by two polymaths, Manfred Clynes and Nathan Kline (1960/1996), they meant a biological feedback mechanism that was not self-aware, attached potentially to human beings who were self-aware, and who thanks to the Cyborg would be more free to engage in thinking, exploring.[4] Cyborgs were planned to be truly indifferent kinds of things, attached to things of an interactive kind. Science fiction modified the word so that cyborgs became self-aware machine-human compounds. These are interactive. The distinction between the interactive and the indifferent holds up surprisingly well for cyborgs, both real and fictional ones. (Hacking 1998c). But I do not count on that situation continuing. George Dyson (1997) may persuade us that we are in the midst of a wholly new stage of evolution, in which artificial intelligence becomes intelligent, in which machines are beginning to participate in evolution itself. Perhaps we will become aware of the ways in which machines are classifying us.

I am not yet obliged to answer questions that arise from this new movement of posthumanism. They are not pressing here, because in none of these questions is there enough of awareness to incline us to talk of interactive kinds. There is feedback, for sure, but not feedback in which self-conscious knowledge plays much of a role. For the purposes of this chapter, devoted to serious mental illnesses that are in our midst right now, we have no need to be futurists. When we get to the future, we will renegotiate our concepts as best we may, in ways that we cannot predict.

Natural and Social Sciences

Although I shall not develop the theme here, I do suggest that a cardinal difference between the traditional natural and social sciences is that the classifications employed in the natural sciences are indifferent kinds, while those employed in the social sciences are mostly interactive kinds. The targets of the natural sciences are stationary. Because of looping effects, the targets of the social sciences are on the move. This is a quite different ground of distinction than others that have been proposed. It has long been insisted that human sciences should not employ methods of the natural sciences, aiming at explanation and prediction, but should try to understand the human agents; *Verstehen* is the watchword in German. That doctrine is intended to replace the largely positivist social sciences of today by human sciences, with different aims and methods. My proposed distinction has nothing to do with that position. The *Verstehen* that enters my story is in the ways in which self-aware people who are the objects of the social sciences may understand how they are classified and rethink themselves accordingly.

PSYCHOPATHOLOGIES

A far more interesting issue is, what happens if something is both an interactive kind and an indifferent kind? Psychopathology furnishes obvious candidates. I do not want to insist on any one psychopathology, but will mention a range of cases. Each of them is to some extent a dreadful mystery, a veritable pit of human ignorance: mental retardation, childhood autism, schizophrenia. It is true that childhood autism was diagnosed only in 1943, and that schizophrenia was named only in 1908, but there is a widespread conviction that these disorders are here to stay, and were with us long before they were named.

There are competing theses about these three examples. One type of thesis tends, speaking very loosely, to the constructionist camp. The other type tends, once again speaking loosely, to the biological camp. In the constructionist camp, these disorders are interactive kinds of illness. In the biological camp, they are thought of as indifferent kinds. Here is a very sharp instance of the fundamental tension between the "real" and the "constructed." I am attempting to address the felt tension with a less tired set of opposites.

We need to make room, especially in the case of our most serious psychopathologies, for both the constructionist and the biologist. That is not to say that I favor one or the other, only that I want spaces in which each can work, without interfering too much with the positive parts of the other's research programs. I shall begin by stating the constructionist attitude to three severe mental disorders.

Biolooping

First, a warning. We have contrasted indifferent and interactive kinds: conscious human beings may interact with interactive kinds of which they are aware. There is a substantially different phenomenon. It too may properly be called interaction, and I want at first to keep it quite separate from the looping effects that I have been discussing.[5]

Everyone knows that our physical states affect our sense of well-being. Many of us believe that our mental states may have some effect on our physical condition. We can learn how to control our nervous tension, or our heartbeat, by a mixture of mental and physical exercises. More crudely, when we think we are ill (or healing) we may become ill (or heal). Changes in our ideas may change our physiological states. Yoga is the technique that spans mind and body most conclusively, and serves as the model for notions of biofeedback. This phenomenon, which is well established but not understood, is distinct from the looping effect of interactive kinds. For lack of better nametags I shall call the mind/body effect *biolooping,* by analogy with biofeedback. The other is *clas-*

sificatory looping. I need the distinction because of course, in particular cases, both types of looping may be at work, and indeed mutually reinforce each other.

Biofeedback usually means a rather conscious control of organic processes, the way in which the master of yoga brings the heart to a virtual standstill. I wish to take in a larger canvas. Oncologists were startled when a team of researchers made a dramatic statistical observation about breast cancer. The prognosis for breast cancer patients who participated in support groups and who had a somewhat optimistic attitude to their illness was, in one large sample, dramatically better than the rate for patients who were resigned, depressed, and who did not work at achieving a better mood. We are talking a (claimed) difference of eighteen months of life added, on average, by mood. (Or is this a selection effect? The positive attitude might be the consequence of having a less virulent tumor in the first place.) This is not because of any mutual awareness between the tumor and the victim. It is true that some psychological approaches to cancer encourage the patient to visualize, become closely in touch with, the tumor, and sense how it reacts to being thought about. Who knows, that might be biofeedback, strictly understood, if it worked. Present studies show no more than that a positive mood and lifestyle are correlated with a better chance of healing or remission. This is not the conscious biofeedback of the yoga master, but it can be called biolooping.

Closer to mental illness, serotonin levels are now correlated with depressive states. An experimental study—not just statistical analysis—is possible. Take a class of patients diagnosed with depression, who improve under purely behavioral treatment. There is no chemical intervention whatever, only a type of psychobehavioral therapy. Results indicate that the serotonin levels of those who improve under such treatment are close to the levels in nondepressive patients, whereas before treatment serotonin was depleted. Once again, for convenience, I shall call this biolooping.

There is every reason to suppose that biolooping and classificatory looping could both be at work in some psychopathologies—and, who knows, in much of ordinary life as well. But first let me focus on classificatory looping. I briefly sketch both the constructionist and the biological attitudes to three terrible mental problems. I start with constructionist visions of each, and then pass to biological ones.

The Feeble Mind

On the construction side we have, for example, *Inventing the Feeble Mind* (Trent 1994), a book that shows how the seemingly inevitable classification, "retarded child," overlaps with and has evolved from a host of earlier labels: ill-balanced, idiots, imbeciles, morons, feeble-minded, mental deficients, moral imbeciles, subnormals, retardates. Each of these classifications has had its moment of glory. The populations singled out overlap markedly. Each label was thought of as a classification or subclassification that improved on previous ones. Each classification has been associated with a regimen of treatment, schooling, exclusion, or inclusion. Each has surely affected the experience both of those so classified and of their families, their schoolmates, their teachers. At various times in our history each classification has been an interactive kind. At the time that each classification was in use, it seemed somewhat inevitable, a perfectly natural way to classify children with various sorts of deficit. Yet when we see the parade of ungainly labels, we quickly realize that these classifications are highly contingent. Each reflects the medical and social attitude of a particular epoch. They could have been otherwise. Chapter 1 stated a background condition (0) for social constructionism: In the present state of affairs, *X* is taken for granted; *X* appears to be inevitable. (0) is well and truly satisfied here. Mental retardation seems like an inevitable concept with which to describe some human beings, but in fact it was an idea waiting for a social-construction thesis to happen to it.

What grade of social constructionism do we reach? Irony and reformism, certainly, but also unmasking: the idea of mental retardation (and all those other names just listed) being part of an ideology whose extra-theoretical function (Mannheim) was to control difficult children, divert them away from schools or school buses into institutions or regimens of treatment. And the retarded have fought back. Every public school in California is required to integrate a certain number of "special education" children into every classroom. This is a splendid example of ideas operating within an extended matrix. One very helpful accident for special education programs was the fact that President John F. Kennedy had a retarded sister, so he set in motion, long ago, federal programs that have ended up as special education in California.

California's programs provide a wonderful illustration of how interactive kinds work. First, the classification has become embedded in a complex matrix of institutions and practices wherein a certain number of children, designated in a certain way, must be assigned to every class, although they are also removed from the class for more individualized tuition. The regular teachers complain bitterly that the result is class disruption; the specially educated know how they are classified; they develop not only individual but collective new patterns of behavior. One can make a strong prediction that not only will the procedures be modified, but also the ways in which these children are classified will be modified because of the new kinds of behavior that have emerged.

These looping patterns also show up in the past. Those changes in terminology referring to retarded children were not the result of a better classification of individuals as pure beings-in-themselves, but reclassification of individuals in the light of how those individuals had altered, in the light of a previous classification and because of the theories, practices, and institutions associated with that classification. One regular refrain in the history of mental retardation is the claim that now we are getting to understand things—as if it were the same thing being understood all along.[6]

Schizophrenia

Or take schizophrenia. Here we have for example *Schizophrenia: A Scientific Delusion?* by Mary Boyle (1990) who, in her preface, avows that she is a social constructionist. Her subject is seemingly less amenable to such treatment than that of mental retardation. Instead of a string of sad and inapt labels for the people classified, moral imbeciles and all the rest, we have only a few neologisms made from Latin or Greek, *schizophrenia* and its precursor *dementia praecox,* and then classifications that no one today has ever heard of, such as *hebephrenia.* Once Eugen Bleuler had given us the name in the first decade of the twentieth century, it stuck.

As Boyle herself says, she concentrates not on schizophrenics but on those who diagnose schizophrenia. She recounts the history of this "kind" of patient. She notes stark mutations in the concept of schizophrenia. She claims that clinicians are often benignly unaware of them. She argues that the introduction, definition, and characterization of this theoretical notion fails to satisfy criteria of adequacy set out by C. G. Hempel, that most careful of logical empiricists. (And tells us, interestingly, how some attempted and influential definitions were devised to deal with points made by Hempel himself.) She argues that psychiatrists, patients, families, welfare agencies, all "need" the idea of schizophrenia. Her conclusion, stated baldly, is that schizophrenia is a construct. Attempts to identify its etiology by neurochemistry are doomed. Schizophrenia is not a kind of disease. The motley of impaired individuals that at different times, and in different ways, have been handily lumped together as schizophrenics are not of a kind.

Schizophrenia, in short, is a scientific delusion. According to my grades of commitment, Boyle is at least a rebellious constructionist about schizophrenia. She wants to unmask and disintegrate. R. D. Laing and other leaders of the anti-psychiatry movement of a quarter-century ago were revolutionary. They were out there in the streets, the clinics, and the wards, trying to destroy and replace this very category of disease.

We need not embrace anti-psychiatry to realize that the classification as schizophrenic, and current attitudes to and treatments of schizophrenics, are matters of which the patients, for all their periodic deficits of logic and sense of reality, are intensely aware. More of them are more aware now than they used to be. This is because of the continually developing arsenal of psychotropic drugs that is already able to bring some semblance of ordinary life to more than half of those patients diagnosed with severe schizophrenia.

The medications make it easier for someone who is afflicted by such a mental illness to think of it as something "other," a thing, almost an agent that acts upon one. One's stupid, or gross, unfeeling, or simply crazy actions can then be blamed on the illness which has become an evil agent. Darin Weinberg (1997), accepting that "constructionist studies demonstrate the profound relevance of social processes to the emergence and assessment of mental disorders in various organizational settings," argues that "mental disorders, once assembled as meaningful objects of discourse and practice . . . exercise their own causal influence" on those who, in a social setting, suffer from the disorder. That would be yet another kind of looping effect.

Classification as schizophrenic affects the sensibilities of those classified in many ways. One of the reasons for the changing symptom profile of schizophrenia is, I suspect, that it is a moving target. There are certain rather widespread phenomena that often lead to the diagnosis of schizophrenia—auditory hallucinations, for example. But even the ways in which people diagnosed as schizophrenic describe these delusive hearings have changed, and the content of the hallucinations has changed. Moreover, the role of hallucinations in the diagnosis of schizophrenia is itself mobile. The founding fathers, Emil Kraepelin and Eugen Bleuler, emphasized above all flat affect, and held that many mental illnesses are accompanied by hallucinations. Just before World War II, Kurt Schneider, intending to operationalize the concept, produced a list of some 12 First Rank Symptoms with auditory hallucinations top of the list. When First Rank Symptoms ruled the diagnostic yard, a lot more people became schizophrenic than would ever have made it in the wards of the Burghölzli hospital during Bleuler's reign.

I conjecture a remarkable looping effect here. Bleuler allowed fairly free expression of auditory hallucinations. They were not important; there were other aspects of one's life to come to grips with. He took hallucinations in stride and paid little heed to them. Hallucinations became ordinary, not to be worried about, neither to be the voice of God to be proud of, nor something to hide from the doctor. Hallucinations became so freely available, unproblematic, that schizophrenics said they had them. So Schneider made them almost a sine qua non of schizophrenia, and yes, they were, at that time. But then as schizophrenia passed from being a disorder that was somewhat in fashion to a diagnosis not wanted any more, flat affect came back, and hallucinations, in the most recent diagnostic manuals, are no longer key. The schizophrenic, as a kind of person, is a moving target, and the classification is an interactive kind.

Childhood Autism

My third example, childhood autism, bridges my first two. The name "autism" was invented by Bleuler to describe a characteristic family of symptoms in the group of schizophrenias. Adult patients lost the usual sense of social relationships, they became withdrawn, gave inappropriate responses, a phenomenon deeply disturbing to family and friends. Then the word "autism" was applied to some children previously regarded as feeble-minded, or even deaf-and-dumb. This was the result of Leo Kanner's many years of study of a quite small number of children. He published it in 1943. At that time the prevailing view, influenced by the (brief!) dominance of psychoanalysis in American psychiatry, was that the autistic child had a "refrigerator mother," one who could not express emotion to the child. This doctrine has by and large passed.

Similar if subtler notions do persist in some schools of psychoanalysis, for example, that of Jacques Lacan, in which childhood autism is still connected with problematic relations between mother and child at a critical stage of maturation.

Cognitive science now rules some roosts. Since autistic children have many linguistic and other deficits, theories of cognition may be invoked. A recent fashion has been to argue that the autistic child lacks a "theory of mind." A single ingenious experiment originally suggested by philosophers has spawned an experimental industry.[7] That is often the case in psychology, where new experimental ideas are as rare and as hard to invent as deep mathematical proofs or truly new magic tricks. But as with retardation and schizophrenia, there continues to be a substantial iconoclastic literature urging that autism is not something people just have, and that autism is no single disorder. Thus we read sentences like this: "Mental retardation is not something you have, like blue eyes or a bad heart" (AAMR 1992, 9). "Autism is the 'way people are' rather than 'a thing people have' " (Donellan and Leary 1995, 46).

Autism may seem problematic for my idea of an interactive kind. Autistic children by definition have severe problems of communication. So how can the classification interact with the children? Part of the answer is that they are in their own ways aware, conscious, reflective, and, in the experience of those who work with autistic children, very good at manipulating other people, despite their problems of lack of affect and rapport. But the example brings out that by interaction I do not mean only the self-conscious reaction of a single individual to how she is classified. I mean the consequences of being so classified for the whole class of individuals and other people with whom they are intimately connected. The autistic family, as we might call it—a family with an autistic child—was severely influenced, and some would say damaged, by the doctrine of the refrigerator mother. The subsequent changes in the family contributed to a rethinking of what childhood autism is—not because one found out more about it, but because the behavior itself changed. Most of the behaviors described by Kanner seem not to exist any more.

Indifferent versus Interactive

There is, then, not only a strong pull towards a constructionist attitude to many mental disorders, but also a great interest in what the classifications do to the individuals classified. One of the defects of social-construction talk is that it suggests a one-way street: society (or some fragment of it) constructs the disorder (and that is a bad thing, because the disorder does not really exist as described, or would not really exist unless so described). By introducing the idea of an interactive kind, I want to make plain that we have a two-way street, or rather a labyrinth of interlocking alleys.

There is obviously another side to this story. There is a deep-seated conviction that retarded children, schizophrenics, and autistic people suffer from one or more fundamental neurological or biochemical problems which will, in the future, be identified. It is not claimed that every person now diagnosed will have the same problem. In the case of schizophrenia, some researchers conjecture that there are at least two distinct disorders, one of which declares itself in late adolescence and is genetic, and another of which may not be inherited. No one maintains that mental retardation is a single disorder, but many believe that specific types of retardation have clear biological causes, to the extent that we can say these disorders simply are biological in nature.

Autism is instructive. There was a debate long ago between the anti-psychiatrist, Thomas Szasz, and Robert Spitzer, who as editor of the Diagnostic and Statistical Manuals has directed American psychiatric nosology since 1974. Szasz argued that MDs should treat only what they know to be diseases. Psychiatrists treat troubled people, but cannot identify any genuine medical conditions, so they should leave the treatment to healers, shamans, priests, counselors. Psychiatry is not a branch of medicine. Spitzer replied: what about childhood autism? We know it *must* be neurological in nature, but we have no idea what the neurology is, so we treat it symptomatically, as psychologists. Is it wrong for us as doctors to try to help autistic children just because we do not yet know the neurology?[8] He took this to be a knockdown argument.

We need not argue that nearly all children diagnosed with autism today have exactly one and the same biological disorder. We need only hold possible that there are a few (possibly just one) basic fundamental biological disorders that produce the symptoms currently classified as autistic. Imagine, however, that there is just one such pathology, call it *P*, and that in reasonable time, we discover what *P* is. A great discovery is reported: "Autism is *P*." Optimists will say that we won't have to wait long. As this book goes to press in July 1998, the International Molecular

Genetic Study of Autism Consortium has just announced the first major linkage of autism to a region on a certain chromosome (IGMAC 1998).

There is a question as to what kind of entity *P* will prove to be. Imagine that in the future it is established that a certain set of genetic markers indicates an inherited biological mechanism producing a certain neurological deficit accompanied by biochemical imbalance. Is the pathology genetic, neurological, or biochemical? It is of no moment, to the present discussion, what sort of thing *P* is. Different hypotheses going the rounds involve a range of genetic, neurological, and biochemical conjectures. Medical science has not properly come to grips with how to classify pathologies which originate in different locations on a gene, work through interaction, produce a family of neurological (or whatever) deficits. There is handwaving in which such a pathology would be called a "biotype." For purposes of the present discussion, I leave the categorization of such conjectured pathologies to future medical science, which will have to negotiate the ways in which they are to be described. Let us posit that there is a pathology *P*, no matter how it will be identified.

By hypothesis the pathology *P* will be an indifferent kind. The neuro-geno-biochemical state *P* is not aware of what we find out. It is not affected simply by the fact that we have found out about it, although of course our new knowledge may, with luck, enable us to intervene and either prevent or ameliorate the pathology. In more traditional jargon, *P* would be a natural kind.

The Bio/Psycho Choice

How can a kind be both interactive and indifferent? My difficulty must be distinguished from two others that are more familiar. Both are of immediate practical importance. First, the issue of *bio/psycho choice.* This is a question of treatment. Even though one may be firmly convinced that a disorder is biological in character, one may realize that the best way to treat it, at present, is psychologically. Classic figures reached this conclusion. Freud never gave up his biological, indeed mechanical, picture of the human mind and its discontents. Every neurosis was biological at bottom. But one could not treat the disorders biologically, and a psychological therapy was needed.

Bleuler is the better example, dedicating himself not to neurotic Viennese but to psychotic Swiss. He was totally committed to the organic basis of mental illness, yet selflessly dedicated to establishing personal and social relationships with schizophrenic patients. At a certain stage in his career, he lived with them night and day; visitors to the Burghölzli were amazed at the ways in which profoundly psychotic patients were able to live in consequence of Blueler's care. He believed in organic psychiatry, but practiced dynamic psychiatry. It helped them heal.[9]

Bleuler opted for psychological treatment, and might well have chosen it even if there were more effective biological treatments available in his day. This is a familiar matter of choice for every psychiatrist. Chemical treatment of the mentally ill is now very much cheaper than psychological treatment. So the pressure for chemical treatment is great, quite aside from the profits that venture capitalists and pharmaceutical companies stand to gain. In ideal circumstances, the bio/psycho choice today is a choice of emphasis rather than rigidity. In the case of depression, some physicians favor writing a prescription for a Prozac-style chemical, and merely monitoring usage. Others favor, for a few patients, purely psychological treatment. Most sensitive practitioners would like to be able to combine the two, using chemicals to ameliorate the worst symptoms, but working on the life issues that provoke unhappiness in the patient. That may be a happy outcome of the problem of bio/psycho choice, but it is a luxury for most clinicians in public service, for they do not have the time for intense psychological care of many clients.

An altogether different type of issue concerns causation. Some mental illnesses are widely believed to result from a basic neurological or biochemical abnormality, typically inherited. It is also thought that they are triggered by some event, possibly organic or possibly social, or possibly socio-organic ("stress"). This is not an exclusively late-twentieth-century view, but one long established in the annals of mental illness. The great neurologist Jean-Martin Charcot was sure that most mental illnesses, including hysteria and epilepsy, were inherited, but especially in the case of hysteria were triggered by life events. There was an ancient formula for expressing the idea: the distinction between predisposing and occasioning causes. Present speculations about the causation of severe mental illnesses such as schizophrenia fit perfectly into that old-fashioned mold: a bio-neuro-genetic predisposing cause, and some occasioning cause, a life problem, an accident, or whatever.

These remarks are an aside, to ward off a confusion I have encountered. The present chapter has nothing to do with bio/psycho choice or the predisposing/occasioning model. Both of course are relevant to the feeling that many kinds of mental illness are interactive kinds, and yet are also indifferent kinds. The clinician who takes a psychological approach may seem to regard an illness as interactive; one who takes a biological (e.g. chemical) approach does seem to regard it as indifferent. If you subscribe to the predisposing/occasioning model of a mental illness, the predisposing cause may be biological, indifferent, while the occasioning cause may be social, interactive. In both these cases, a tension is apparent, one of great importance in each case. But I am worrying at a different source of tension, more of a logical dilemma than a medical or clinical one.

A DILEMMA

Suppose that childhood autism is at bottom a biological pathology *P*, namely what has traditionally been called a "natural" kind and what I here call an indifferent kind. What then happens to the claim that childhood autism is an interactive kind? That is, a kind in which the humans classified may indeed change through looping effects, because of the ways in which the people classified react to being so classified? How can it be an interactive kind and also an indifferent kind?

This is one way in which to address an issue that troubles many cautious people, the idea that something can apparently be both socially constructed and yet "real." This is quite distinct from my clumsy attempt to argue that child abuse is both socially constructed and real. For there we can make a trifling distinction. The idea of child abuse (and the entire surrounding matrix) is socially constructed, while child abuse is real. Here we want to say both that childhood autism *is* (is identical to) a certain biological pathology *P*, and so is a "natural" kind or an indifferent kind. At the same time, we want to say that childhood autism is an interactive kind, interacting with autistic children, evolving and changing as the children change.

The pathology *P* causes havoc in the behavior, life, and emotions of conscious, judging, moral, aware, somewhat autonomous human beings, namely autistic children. But pathology *P* is, by hypothesis, not what it is in virtue of anything conscious, self-aware. The greater the role of fundamental genetics, of molecular identification in the pathology *P*, the more people say that human genome is the place to look, then the more obvious it will seem that we are in the realm of indifferent, "natural" kinds.

Semantic Resolution

At this juncture, philosophers may like to think of childhood autism and the postulated pathology *P* in terms of the theories of reference advocated by Hilary Putnam (1975) and Saul Kripke (1980). The term "autism" is what they would call a natural-kind-term, analogous to the multiple sclerosis that Putnam long used as an example (even before working out his theory on the meaning of "meaning.")[10] If there is in fact exactly one definite biological pathology *P* underlying a broad class of autistic children, then the reference of the name "childhood autism" is *P*. Under this hypothesis, the name "childhood autism" is, in Kripke's terms, a rigid designator of a natural kind, namely the pathology *P*. In my terms, the pathology *P* is an indifferent kind, and "childhood autism" is the name of that kind.

Our difficulty then seems merely verbal. Yes, if there is precisely one neuropathology *P* underlying what we now call autism, then, in Kripke-Putnam semantics, the kind-term "childhood autism" rigidly designates that pathology. Shall we say that when Kanner coined the name "childhood autism," it referred to pathology *P*? Some would give him what Putnam calls the "benefit of the dubbed"—yes, he referred to *P*, even though he (like ourselves) had not the remotest idea what childhood autism really is, namely *P*.

Putnam's theory of meaning presents meaning as a vector, or ordered tuple. This vector is in most ways like a dictionary entry: part of speech, category, down through stereotype, but ending in an item no dictionary, or anything else, can ever present: the extension of the term being defined. That is, the class of things falling under the term, the class of things to which the term applies. In our example, the final entry in the meaning of "meaning" vector for "autism" is the pathology *P*, or perhaps all instances of the pathology *P*.

We can perfectly well keep Putnam's machinery, but suppose that in the Putnam-style meaning of "autism" (and of a great many other words) we put an enriched stereotype of childhood autism, the current idea of childhood autism, accompanied by definite examples and descriptions of prototypical autistic children. So-called definitions of mental disorders commonly proceed by giving clinical examples prototypes. We need not now concern ourselves with details. In the vector for the meaning of "childhood autism" we should include both the current idea of autism—prototypes, theories, hypotheses, therapies, attitudes, the lot—and the reference, if there is one, namely the pathology *P*.

Now for the bottom line. Someone writes a paper titled "The Social Construction of Childhood Autism." The author could perfectly well maintain (a) there is probably a definite unknown neuropathology *P* that is the cause of prototypical and most other examples of what we now call childhood autism; (b) the idea of childhood autism is a social construct that interacts not only with therapists and psychiatrists in their treatments, but also interacts with autistic children themselves, who find the current mode of being autistic a way for themselves to be.

In this case we have several values for the *X* in the social construction of *X* = childhood autism: (a) the idea of childhood autism, and what that involves; (b) autistic children, actual human beings, whose way of being is in part constructed. But not (c) the neuropathology *P*, which, ex-hypothesi, we are treating as an indifferent kind, and which Putnam would call a natural kind. A follower of Kripke might call *P* the essence of autism. For us, the interest would be not in the semantics but the dynamics. How would the discovery of *P* affect how autistic children and their families conceive of themselves; how would it affect their behavior? What would be the looping affect on the stereotype of autistic children? Which children, formerly classified as autistic, would now be excluded, and what would that do to them?

What if there is no pathology *P*, or no P_1, P_2, and P_3? Childhood autism continues to be a good example. One author, who describes herself as a "recovered autistic," distinguishes "autism subtypes" and writes that "The subtypes are on a continuum that merges together" (Grandin n.d.). At one end of the continuum we have Kanner-Asperger Type (high-functioning). Kanner is the physician who gave us child autism. At the other is "Regressive/Epileptic Type (Often Low Functioning) (Late-onset children often lose speech between 18 months to three years old)." Anyone who reads even slightly nonorthodox accounts of autism may well suppose we do not have a linear continuum at all, but an extremely dense manifold of problems, and perhaps not even a set of pathologies.

Or some set of sets of neuropathologies, and a lot of developmental history required to produce any individual case.

Any such scenario makes the Kripke-Putnam semantics seem somewhat irrelevant. I imagine that with the constant thrust towards the biologization and indeed genetization of mental disorders, we shall find that the case I have to some extent imagined is in fact the norm. Semanticists may derive interesting formulations for the new situation. Students of semantics who dislike Kripke's approach will say, if we cannot use a rigid designation for childhood autism, and yet childhood autism is an important concept, why do we need a rigid designation for the meaning of multiple sclerosis either?

My position here is rather curious. I have already made amply plain that I do not, myself, favor the language of social construction. I am discussing it in connection with psychopathologies because many deeply committed critics of psychiatric establishments find social-construction talk helpful. It enables them to begin with a critique of practices about which they are deeply skeptical. I respect their concerns, and have, I hope, represented them fairly, if cautiously. On the other hand, I also respect the biological program of research into the most troubling of psychiatric disorders. That creates a dilemma.

I have suggested a semantic way for a philosopher to make peace with the dilemma. Some would say that it is more than that—it is a tidy resolution of the dilemma. But not only am I ambivalent, or worse, about social construction; I am also ambivalent about the use of rigid designation in connection with disease and disorder. Some of these qualms were well stated quite a long time ago by Avishai Margalit (1979). He wrote when enthusiasm for the Putnam-Kripke approach was at its peak. He argued that even in the case of quite well understood afflictions, such as multiple sclerosis, there are many problems about taking the model very strictly.

Even if Margalit's criticisms are compelling, semantical theories like those of Kripke and Putnam are not rendered useless. They are tools. A screwdriver is not the worse for being a bad hammer. Semantical theories are not literally correct descriptions of natural language. They are artificial ways of construing natural languages for this or that purpose. I do think that these philosophical theories are wonderfully suitable for diverse purposes.[11] In the present case, putting a theory of reference alongside social construction shows how to diminish a felt dilemma. If this approach helps, then it does a real service, for it enables us to move on to more significant issues, to what I call the dynamics, rather than the semantics, of classification.

For the Study of Dynamics, Not Semantics

In the end, the "real vs construction" tension turns out to be a relatively minor technical matter. How to devise a plausible semantics for a problematic class of kind terms? Terms for interactive kinds apply to human beings and their behavior. They interact with the people classified by them. They are kind-terms that exhibit a looping effect, that is, that have to be revised because the people classified in a certain way change in response to being classified. On the other hand, some of these interactive kinds may pick out genuine causal properties, biological kinds, which, like all indifferent kinds, are unaffected, as kinds, by what we know about them. The semantics of Kripke and Putnam can be used to give a formal gloss to this phenomenon.

Far more decisive than semantics is the dynamics of interactive kinds. The vast bulk of constructionist writing has examined the dynamics of this or that classification and the human beings that are classified by it. Studies of Authorship, Brotherhood, the Child Viewer of Television, and Danger have, in their various ways, been concerned with just that: the social construction of the idea of *X*, of *X*, of the experience of being *X*, and so on, and how these interact with each other. Is there anything to say in general about such dynamics, over and above particular and idiosyncratic examples? How does the making and molding of an interactive kind, be it child abuse or autism, help to make up people? How do people make themselves up, as they act in ways that conform to, or stay away from, powerful classifications?

For a compelling example, take biolooping. A person undertakes a certain regimen of behavioral modification, intended to diminish the symptoms and feelings of depression. Numerous kinds of behavior are reinforced, all of which run counter to the classification *depressed.* The patient starts to live in this new way. If the behavior modification works, then even our psychiatric understanding of depression changes. Yet simultaneously, by living in this way, adopting certain types of behavior, a certain chemical condition of the brain, thought to be correlated with depression, is alleviated. We have a dynamics working at the level of classification and at the level of biolooping.

Semantics intrigues the logician, but the dynamics of classification is where the action is. If we begin to move among cyborgs, or to become cyborgs, biolooping will become a common fact of everyday life. Classificatory looping will continue alongside it until, perhaps, the two become one in a world that no one can foresee.

Chapter Five

KIND-MAKING: THE CASE OF CHILD ABUSE

In Chapter 1 I severely criticized my own statement that child abuse "*is* a real evil, and it was so before the concept was constructed. It was nevertheless constructed. Neither reality nor construction should be in question." The statement confuses the real evil, child abuse (real behavior, real events), with the concept, the idea, of child abuse.

Did my statement merit the rebuke? Naomi Scheman (1997, cf. forthcoming) has said that everything that is socially constructed is real. That sounds right. If something has in fact been constructed, then it does exist, and so is real! So child abuse, if constructed, is real, and there was no confusion after all. But Scheman's comment is not as straightforward as it seems. To see why, let us use two examples, one of child abuse, which is real, and the other a recent scare, which, arguably, is only a construct.

"The idea of child abuse is socially constructed." Any idea that is debated, assessed, applied, and developed, is situated in a social setting. It is therefore vacuous to say that every idea is constructed. No point is served by saying that the idea of digging (a ditch, say) is constructed. But there is a point in saying that the idea of child abuse was constructed, or, as I prefer to say, made and molded. The explicit idea emerged at a definite time (1961) at a definite place (Denver) in the discussions of some authoritative people (pediatricians). The immediate reference was battered babies, but the reference was very quickly extended. New connotations were acquired. The idea became embedded in new legislation, incorporated in practices, and changed a wide range of professional activities involving social workers, police, schoolteachers, parents, busybodies. It acquired new moral weight: child abuse became the worst possible vice. If someone wishes to call this a story of social construction, fine. The point of that statement is clear enough.

"Child abuse is real." The point of this statement is also clear, although the statement was more salient in 1962, when few people believed that child abuse was at all common, than it is today, when authorities speak of an epidemic of child abuse. (There is seldom much point is saying something that almost everyone knows is true.) Child abuse is not something that has been imagined by activists; there are innumerable cases of children who have been physically, sexually, or emotionally abused. *That* is the point of saying that child abuse is real. Bringing that fact to the attention of the public, of parents, of teachers, of legislators, and of the victims themselves was one of the most valuable pieces of consciousness-raising to take place between 1960 and 1990.

Contrast sadistic satanic ritual child sexual abuse, SRA (satanic ritual abuse) for short. Undoubtedly the idea of SRA is a real idea, perhaps a very bad idea. There is an interesting history, some of it told in Showalter (1997), of how this idea came into being and was greeted with almost hysterical enthusiasm. Reading her, and other historians, one may want to conclude: "The idea of SRA was socially constructed." That is, there is a specific history of the formation and circulation of this idea, the power that it exercised over a substantial number of people, the practices of suspicion and surveillance that it led to, the police investigations, the commissions of inquiry.

"SRA (unlike child abuse) is not real." That, in some quarters, is a very controversial statement. Some people believe that incidents of sadistic satanic ritual child sexual abuse are not only real but common. My question here is not whether the statement is true, but—what is the purpose of making the statement?

About 1990, in Great Britain, there was a wave of reported cases of SRA which became a national scare. This led to the most systematic and exhaustive official investigation of SRA that has been conducted anywhere. It took six years. Every public allegation of SRA in the United Kingdom was examined. The commission, chaired by Jean La Fontaine (1998), found that none of the charges was substantiated by any evidence whatsoever. What about the past? The worst all-time sexual abuser of children (he liked to bugger small boys and girls before and after torturing them to death) was Gilles de Rais (1404–1440). He was tried explicitly to determine the satanic content of his murders, but even the inquisitors, the best trained and most diligent investigators of satanism the world will ever know, could not bring in a conviction for satanic-ritual-child-sex-abuse although both civil and ecclesiastical courts found him guilty, after torture, of endless other heinous crimes, and he was duly hung.[1]

La Fontaine in no way minimizes the incidence of child abuse. How could she, when just as her book was published, the most extensive British report, based on longitudinal interviews with young adult women, produced statistics on the high range of incidence and prevalence of child abuse that the subjects had experienced in childhood or very early adolescence? A reader of La Fontaine, the longitudinal study, and the case of Gilles de Rais, might sum up the conclusions as follows: "SRA, at least in Britain, is a fantasy. We conclude in the light of the available evidence that there have been no real cases of SRA in the United Kingdom. In contrast, child abuse—physical, emotional, and sexual—is all too prevalent. Child abuse is real. We also have every reason to think that there were a great many real cases of child abuse in our past as well—but precious little evidence of SRA." That is the point of saying: "SRA (unlike child abuse) is not real."

I have published three essays on child abuse. The first, "The Making and Molding of Child Abuse" (Hacking 1991a) is fairly easily accessible. The third, a chapter in *Rewriting the Soul* (Hacking 1995b, ch. 4) is also quite easily found. The second essay (Hacking 1992a) is, however, hard to obtain, and so has been used as the basis for this chapter. Each of these studies was, when published, up-to-date. I wanted to give the sense of a powerful idea being molded before our very eyes. I have not attempted to update this chapter. When writing it I wanted to emphasize how we have concepts, practices, institutions, and even people being formed and molded before our very lives. Philosophers do have a habit of examining very old ideas. Hence Hegel's tag about the owl of Minerva flying at dusk; philosophy takes off after the day is done. I wanted to do philosophical analysis, if not in the heat of the midday sun, at least in the early afternoon. There is no point in repeating that exercise, for of course any updating of the topic of child abuse would at once proceed to become out of date. Instead this chapter now gives real meaning to my talk, in earlier chapters, about an idea being molded in a matrix of very different types of elements.

In revising this essay I aimed at illustrating the complexity of what must by now be a classic case of what is called social construction.

Classic? Child abuse has been called a "social construction" since Gelles (1975). There have been ironic papers on the discovery of child abuse (Pfohl 1977) and, lest we miss the irony, scare quotes are used, as in a paper on the "discovery" of sexual abuse (Weisberg 1984). The song goes on. After the original publication of my essay, Janko wrote a book subtitled *The Social Construction of Child Abuse* (Janko 1994), and Marshall wrote a thesis titled *The Social Construction of Child Neglect* (1993).

RELEVANT KINDS

The American philosopher Nelson Goodman (1906–98), as quoted in Chapter 2, said that he has a *constructionalist* orientation—a word of his own devising. In 1978 he published a set of lectures under the title *Ways of Worldmaking.* "Kind-making" is no empty add-on to worldmaking, for Goodman had a lot to say about kinds, classes, sorts, types. His doctoral dissertation of 1940, *A Study of Qualities* (Goodman 1990), was about some basic kinds; it was motivated by the idea that "without some techniques for applying symbolic logic to extra-logical subject matter, problems that require symbolic logic will never yield clear and precise solutions." He thereby started on the road back to an earlier vision of logic, where a proper study of mankind is kinds. What kinds? Relevant kinds. "I say 'relevant' rather than 'natural' for two reasons: first, 'natural' is an inapt term to cover not only biological species but such artificial kinds as musical works, psychological experiments and types of machinery; and second, 'natural' suggests some absolute or psychological priority, while the kinds in question are rather habitual or devised for a purpose" (Goodman 1978, 10). Kinds are the core of Goodman's philosophy. Perhaps his first completely original discovery was "the new riddle of induction" (Goodman 1954/1983, 72–80). It shows that whenever we reach any general conclusion on the basis of evidence about its instances, we could, using the same rules of inference, but with different preferences in classification, reach an opposite conclusion.

Many of the more logically minded among his readers think that the new riddle is a trick: it should be defeated by definitions and distinctions. His wide readership among social scientists, literary theorists, and students of aesthetics tends to ignore it as technical juvenilia, having little to do with Goodman's larger concerns. Both groups are wrong.[2] There is no general solution to his new riddle. Its scope goes far beyond induction and other trifling modes of reason. It confirms his doctrine, admired in some quarters, detested in others, that we can and do inhabit many worlds. It underwrites his enduring conviction that "without the organization, the selection of relevant kinds, effected by evolving tradition, there is no rightness or wrongness of categorization, no validity or invalidity of inductive inference, no fair or unfair sampling, and no uniformity or disparity among samples" (Goodman 1978, 138–39).

There is a certain ambiguity in the idea of selecting and organizing kinds. Is that something that an individual can do, or is it something essentially social and collective? Both. One example of largely individual selection and organization is the way in which Sudden Infant Death Syndrome became, largely through its selection and organization by a handful of people, an essential element in the portfolio of pediatric and social problems. But since Goodman speaks of evolving tradition, he must have had something more communal in mind. This selection and organizing must have close affinities to what is called social construction. A precondition for reasoning, in a community, is that by and large classifications are in place and shared, although they can also always be invented and modified. The selection and organization of kinds determines, according to Goodman, what we call the world—although he thinks we are better off without a concept of the world at all. The world well lost, as he once put it.

Very well. But how do we do all the amazing tricks implied by Goodman, right up to the making of a world? How do we organize and select relevant kinds? Or should we use the passive tense, how do they come into being? Goodman writes of a "fit with practice" that is "effected by evolving tradition." But he tells us precious little about practice or evolution. I want to begin to fill the gap, giving just one example of the complex ways in which a kind can be made and molded—and, in the end, change the world.

There can probably be no general theory about selecting kinds. There are many types of kinds, and no one has done more than Goodman to remind us of this. Yet although he regularly writes of "motley entities," even he tends to put all kinds into one basket, precisely to de-emphasize absolute priorities and to emphasize that artificial kinds are as important to us as the kinds of things we find in nature. There is no harm in using one big basket tagged "relevant kinds." A basket is not a food-processor that annuls difference. The basket of the harvest festival is bountiful just because it displays so variegated a collection of grains and flowers and fruits and vegetables. Yet putting things in the basket does make one see them in one way, as harvest, as bounty, as worthy of thanksgiving, for example. I want to take some out of the basket again, and see them in different ways. This chapter will look at only one example, to show what a rich, varied, and confusing mass of material lies under that agreeable euphemism, "the selection of relevant kinds."

We need a detailed example to get some sense of how, in ordinary life, we select and organize new kinds. We need an example of evolving tradition, not evolution over a thousand years, but evolution over a few decades. "Child abuse" serves. It is both human and scientific. Scientific? Let us mean, for present purposes, whatever aims at science, what passes as science, what models itself on the methods of established and successful science, what claims to discover objective truth about the world and its inhabitants, what claims to give explanations, to make falsifiable conjectures, to increase our power to predict, control, and improve. A kind can be embedded in science thus understood, without being indifferent or "natural." No one is astonished to read that "child abuse is not a naturalistic category—nothing is 'naturally' child abuse" (Parton 1985, 149). Despite a steady thrust to medicalize it—many authors speak of the "medical model" of child abuse—the semantics of child abuse is not in the least like the semantics of schizophrenia or autism suggested in Chapter 4. There is no underlying pathology to be discovered, which is uniquely associated with a propensity to abuse children, and such that a major segment of the population of child abusers have that pathology.

Child abuse is an interactive kind. Interactive kinds interact with people and their behavior. I conclude this chapter with something even harder to understand. We can well understand how new kinds create new possibilities for choice and action. But the past, of course, is fixed! Not so. As Goodman would put it, if new kinds are selected, then the past can occur in a new world. Events in a life can now be seen as events of a new kind, a kind that may not have been conceptualized when the event was experienced or the act performed. What we experienced becomes recollected anew, and thought in terms that could not have been thought at the time. Experiences are not only redescribed; they are refelt. This adds remarkable depth to Goodman's vision of world-making by kind-making.

There are less novel ways in which some interactive kinds differ from most indifferent kinds. Many of our sortings of people are evaluative.

But surely not scientific sortings, sortings of medicine or the positive social sciences? Yes, those too. Many of the kinds that have emerged in social science are kinds of deviance, typically of interest because it is undesirable for the person to be of that kind. Such social sciences aim at providing information to help people in trouble. Classifications evaluate who is troubling or in trouble. Hence they present value-laden kinds, things to do or not to do. Kinds of people to be or not be. Partly because of implied values, people sorted under those kinds change or work back upon the kind. Most of us want to be seen as good and confess our sins as bad. Socrates argued that all human beings seek the good, and there is something to be said in favor of his complex paradox.

Classifications can change our evaluations of our personal worth, of the moral kind of person that we are. Sometimes this means that people passively accept what experts say about them, and see themselves in that light. But feedback can direct itself in many ways. We well know the rebellions of the sorted. A classification imposed from above is rearranged by the people to whom it was supposed to apply. Gay liberation is only the most successful example of this type of interaction. Interactive kinds are involved in "making up people." There is no single story to be told about that. One gets a grip on how a kind works only by studying it in some depth. A study of one kind may illuminate many others. But no matter how well chosen the example, it will serve only as a guide for understanding a group of kinds. It should never aim at being a model for all kinds. The motto is "motley."

WHY THIS KIND?

We can think about kinds using simple and stylized examples or using ones that are complex and lively. Timeless abstract illustrations have a central place in philosophy, whose proper study is the right and the good. But great generalities and abstractions should not transfix us. We also need to examine concepts that are highly specific, current, and dense. Child abuse as a kind of human behavior, and its correlates, child-abuser, and abused child, are dense indeed.[3] To use abuse as a philosophical example may seem to imply a lack of sensitivity that verges on immorality. It distances author and reader from pain and victims. I shall not defend the choice by the priggish claim that the idea of child abuse could do with some philosophical analysis. There is indeed an immense amount of conceptual confusion about the idea. Reading the professional literature fills one with gloom, not only about the fates of children but also about ritually institutionalized research and writing. But a philosophical study of kinds will not change that. My reasons for choosing the example are practical. We can watch it. It is happening right now.

Thus a first reason for picking child abuse is to study a current and pressing kind. It is not so often that we can experience a concept in rapid motion. Goodman spoke of evolution but gave little hint of what evolves and how. We shall see more than evolution in three decades of child abuse; we shall see mutations worth calling revolutions, conceptual displacements worth calling explosions.

A second reason for choosing the example is that child abuse is a "relevant" kind in more senses than Goodman's. The selection of child abuse as a vital classification has had enormous consequences in the law, in day to day social work, in policing the family, in the lives of children, and in the way in which children and adults represent their actions, their past, and those of their neighbors.

Third, despite its role in social rhetoric and politics of numerous stripes, child abuse was first presented and is still intended to be a scientific concept. There are demarcation disputes for sure: is this a topic for medicine, psychiatry, sociology, psychology, jurisprudence, or self-help? Whatever the standpoint, there are plenty of experts firmly convinced that there are important truths about child abuse. Research and experiment should reveal them. We hope that cause and effect will become better understood, so that we can find predictors of future abuse, that we can explain it, that we can prevent it, that we can determine its consequences and counteract them. We hope that we can cure child-abusers and heal hurt children.

A fourth reason for picking child abuse is that for all that we would like an objective concept about which expert knowledge is possible, the idea of harming innocent children is powerfully moral. In our present system of values, genocide is the worst thing that one group of people can do to another. Abusing a child is the worst thing that one person can do to another. We cannot have a better example of a scientific kind that is also a moral one.

My orientation is Goodman's, skeptical and analytic. There are notes of criticism and irony in my sketches below. I may seem to be in the business of reforming, of showing what is wrong with the child-abuse movement or social science in general. That is not my intention. My interest is rather in the way which "child abuse" and "abuser" and "abused child" denote kinds, and what those kinds do to us. They differ, in numerous ways, from exemplary kinds in the natural sciences. That does not prove that they are not "scientific." I should also say that the fascination with the idea of child abuse, as I sketch it below, does force some hard questions. For example, "for all its horror, child sexual abuse (or physical battering) harms, indeed kills, far fewer children, either in [the United Kingdom] or the United States, than simple, miserable and unremitting poverty. Why, when poverty has been intensifying and welfare programmes run down, has our attention been drawn to sexual or other abuse?" (Beard 1990). The author thinks that part of the answer is that child abuse offers scapegoats: "for anyone who sees poverty and deprivation as the bigger enemy, single-minded preoccupation with sexual abuse must seem a dangerous deflection." That was also the message of the most serious study of deaths due to abuse and neglect. The children who die from maltreatment are the poor ones (Greenland 1988). Yet the classification that has been constructed has been quite deliberately made as far away from poverty and welfare as could be.

In the United States, where so much of my story takes place, the availability of public funds for poor families with small children decreased substantially every year in the decade after 1981, while every year we heard more and more about the horrors of child abuse, culminating in 1990 with the statement by a Presidential Panel that it was a "national emergency."[4] The board said that its first tasks were to "alert the nation to the existence of the problem." Then what? "We want a system in which it is as easy for a family member to get help as it is to report a neighbor for suspected abuse." But don't bring up unpleasant topics like filth, danger, and the stench of urine in the halls, elevators that don't work, smashed glass everywhere, cancellation of food programs. Just tell us that your dad is abusing your little sister.

A SKETCH OF HISTORY

Cruelty

"Child abuse" as a way to describe and classify actions and behavior came into being in discussions and observations that took place in Denver, Colorado, around 1960, and first went public at a meeting of the

American Medical Association in 1961. That sounds paradoxical. Is not child abuse just cruelty to children under another name? No. I shall only summarize the argument here.[5]

Child abuse emerged as one of the earliest socio-political causes in the 1960s, although it became truly radical only at the end of the decade. Cruelty to children was one of the last of the great Victorian crusades, and came after anti-slavery, factory legislation about child employment, temperance, the extension of the suffrage, anti-vivisection, and cruelty to animals. The first formal organization dedicated to fighting cruelty to children was the New York Society for the Prevention of Cruelty to Children, established in 1874, as an adjunct to the Humane Society, whose job was to prevent cruelty to animals. Cruelty to children was never radicalized. Here I speak, of course, of the institutional awareness of cruelty. Creative artists tend to be decades and even centuries ahead of do-gooders. There is no more powerful condemnation of violence against children than etching no. 25 in Goya's series of 1799, *Los Caprichos.* A child, buttocks bared, is being beaten furiously by a crone with a shoe. Title: "But he broke the pitcher" *(Si quebró el cántaro).*

Many instances of what Victorians called cruelty to children we now call child abuse, and vice versa. But the two types of classification of behavior are not identical. There are, indeed, plenty of analogies. When we examine a larger scene we see many resemblances between populist or charitable reform leagues in the 1880s and those begun in the 1960s. Some seem to be repeats, even down to details such as the enthusiasm on the part of some groups for forcibly separating parents and children. Women's organizations have comparably focal roles in both periods. There are nevertheless very general grounds of difference that we can quickly enumerate.

First there is the matter of social class. Cruelty to children, in the Victorian mind, was a matter of poor people hurting their children. Child abuse, as it emerged in America in 1960s, was deliberately presented as classless, as equally common in all social classes. Why? In order to form a broad political front; in order that child abuse should not be seen as an exclusively liberal, social-reform type of issue.

Second, the Victorian activists loathed cruelty to children but were not frightened by it. Risk was not a word in play then; it was central, however, for the rhetoric of the 1960s. Mary Douglas and Aaron Wildavsky (1986) have argued that risk and pollution very often go hand in hand. We can hardly have a better example than the case of child abuse.

Abuse is not only an ultimate evil, but also an ultimate pollution of the child, of the family, of the society. "Children at risk" has become a virtual catchphrase. Cruelty to children did not involve much talk of risk or of pollution. Cruelty to children was bad. But it was not an ultimate evil, inducing thoughts of horror and disgust.[6]

Third, cruelty to children was not a medical problem, while child abuse was medicalized from the beginning. The idea was brought forward by pediatricians. Child abusers were described as ill. Medicine has by no means kept uniform control of the administration of child abuse, but whoever aims at control must treat child abuse within some science. Contrast cruelty which was not scientized. Men who beat their children were not subject to medical scrutiny as a special kind of sick person. A great deal of what Donzelot (1979) calls the "the policing of families" made use of medical theories, but cruelty followed another route. People did not try to control it by means of a special kind of knowledge about the cruel. They never made out that the cruel parent was a "kind" of human being about which specialized knowledge was possible.

Fourth, Victorian courts had plenty of cases of sexual offenses against children, but sexual assault or seduction was not categorized as cruelty to children. Not only were the offenses dealt with under other statutes, but the discourse of the day simply did not link cruelty to children either with sexual assaults on children or with their seduction.

Cruelty to children faded from notice by 1910. The Children's Division of the American Humane Association, and similar state and national organizations, were maintained but all were to some extent displaced by an emerging profession. The very name "social worker" was virtually unknown before 1900, yet by 1910 there were schools of social work in several nations, starting with the Netherlands. By 1912 there was a flourishing National Social Workers Exchange in the United States, with an Employment Bureau listing a sizeable number of categories, properly including family work, broken homes and neglected children. The older tradition of charitable amateurs was dismantled or reorganized with professionals on the streets or in the courts. In the fifty years 1910–1960 there were plenty of problems about children and adolescents. Infant mortality and juvenile delinquency took their places ahead of cruelty to children.

American attention can be charted even from the programs of the successive White House Conferences on children, inaugurated by President Hoover, who gave the immortal advice that the nation must now attend as carefully to the child crop as to the farm crop. This analogy was powerful at least until 1941. A standard fable of the Midwest (Indiana, in this case) has a sick and single mother of hungry infants begging uselessly for help from state and federal authorities. But a wire to the United States Department of Agriculture about a case of "hog cholry" gets the reply, "Cert, I'll send you a man right away." The story ends, "Anybody, even a fool, can see it would be cranky for the state to save the life of a little mother, and could not afford it either. MORAL: Be a hog and worth saving." (Goddard 1927, 195–7). Child mortality, child health, and adolescent crime were the prime issues for half a century.

Then, in 1961–62, came child abuse. The immediate stimulus came from a group of pediatricians in Denver led by C. H. Kempe. They drew attention to repeated injuries to small children. X-rays were the objective proof. Children were found to have healed fractures in legs or arms, and similar signs of unrecorded, unreported injury. There had even been a "syndrome" in the roentgenographic (X-ray) literature, which no one had dared to say had been caused by parents beating up their babies. One would hardly have guessed the topic of the 1945 paper, "Infantile Cortical Hyperostos: Preliminary Report on a New Syndrome." Only in 1961 did the Denver group announce "the battered child syndrome." They published in 1962 with the full majesty of the American Medical Association behind them. Newspapers, television, and the mass weeklies announced this new scourge. In 1965 the *Index Medicus* added child abuse to its list of medical categories to be catalogued. General-interest indices such as that for the *New York Times,* which had previously had the listing "cruelty to children," began to run two entries, the second being "child abuse," which then became the standard place to file the stories on abuse and cruelty. Meanwhile, "Kempe" became the name not just of a man but of a radical break in our awareness, to the extent that soon one could write retrospectively about "Child Abuse before Kempe" (Lynch 1986).

It is striking how some of the "knowledge" about child abuse was asserted from the start as part of the conceptual, analytic frame of this newly noticed kind of human behavior: "Often parents may be repeating the type of child care practiced on them in childhood" (Kempe et al. 1962, 23). Soon we had: "abused as a child, abusive as a parent." That became an item of belief in America by the well-intentioned, moderately informed, liberal-leaning population at large. The literature on the "inheritance" of child abuse was nevertheless incredibly mixed, with firm believers ranged against complete sceptics.

The believers held the field, and for two reasons. First, the claim sounds right, that is, it fits in with twentieth-century beliefs about childhood experience forming the adult. Secondly, it is now a foregone conclusion that an abusive parent will profess being abused as a child. That explains and thereby mitigates the behavior. So there is plenty of "confirming" evidence. I am not saying that the proposition is false. I am saying that the grounds for accepting the proposition as true had little to do with evidence. The statistical studies on both sides exemplify the role of statistical technology in the legitimation of the passions—not "garbage in, garbage out," but "beliefs in, beliefs out." By 1995 there were over ninety major statistical studies. With such a wealth of data, it is usually possible to do statistical meta-analysis that detects underlying patterns even when individual studies seem to be in conflict. That has not yet been the case with child abuse.[7]

Also from the beginning the remedial agenda included the practical injunction to separate babies from their parents or caretakers: "Physicians should not be satisfied to return the child to an environment where even a moderate risk of repetition exists" (Kempe 1962). The entire topic was declared in need of medical expertise: "It is the responsibility of the medical profession to assume leadership in this field" (Helfer 1968, 25). The popular press was faithful, speaking of "sick adults who commit such crimes." There are two points here. First, that there is a knowledge, a truth about child abuse to be had, and, secondly, it is the doctors who should have it. Even those who protest against medical control seem unable to escape the medical conceptualization. For example, two sociologists, in an unusually wise study of the idea of child abuse, note that "child abuse" has come to denote far too many things. "Commonality is yet to be demonstrated in the diverse phenomena that are considered to be manifestations of abuse and neglect." We need "greater specificity in policy-related research endeavors, including epidemiological, etiological and evaluative research. Until there is further delineation of that which is to be counted and estimates of its dispersion, epidemiological and incidence estimation would seem to be futile. Similarly etiological research may be premature . . ." (Giovannoni and Becerra 1979, 239). Note how this book of sociology has to express its ideas in medical Latin, etiology and epidemiology.

The fusing of events with little "commonality" made it easy to create a popular front. The resulting political program was brilliantly described by Barbara Nelson (1984), in the tradition of Joseph Gusfield's work on the setting of socio-political agendas. Whereas in 1962 there was no specific legislation anywhere in the world for the reporting of battered children, a host of laws and agencies soon sprung into being at the national, state, and local level, first in the United States, then in other parts of the English-speaking world, and then in continental Europe. Because of the individualist climate of opinion peculiar to the United States, it was an essential part of the American political agenda to separate the problem of injured children from any social issues. "This is a political problem, not a poverty problem" insisted Senator and then Vice President Mondale, who led the drive for national legislation. Liberals and conservatives could agree on something, so long as social issues did not arise. If child abuse is a sickness, then we can act in unison to combat it. President Nixon signed the Act dealing with child abuse in 1974. Only one voice in the Senate was opposed (Jesse Unruh).

Sex

Battered child syndrome applied to babies three years old and under. The Denver pediatricians said later that they had made the deliberate decision not to go public with "physical abuse" as a general label for what was happening in many American families. They feared that a conservative audience of colleagues would not put up with it. But once searing photographs of damaged innocents were in place—injured not only with sticks and stones but by straps, nails, cigarette butts, scalding water—it would quickly be acknowledged that babies are not the only victims. It had, however, helped to begin with infants, where issues of punishment and the authority of the parent could be evaded. Babies are, in our scheme of things, too young to be punished, let alone with such brutality.

Once the cry was raised, battered babies would be seen as only a subclass of the "real" kind, the abused child. Sacrosanct privilege could be challenged. Was not corporal punishment a species of child abuse? Families and schoolteachers could be policed to ensure that they did not beat children. There were nonphysical variations on the theme, such as "confinement abuse," tying children to bedposts for days, or locking them away in dark closets. But in the sixties child-abuse-and-neglect meant physical abuse and neglect. Sex was peripheral or absent. The pioneers of 1962 said later that they were well aware of sexual abuse, and had it on the list of future targets. Police officers, social workers, psychotherapists, and ministers of religion certainly knew that sexual abuse and physical abuse often occur in the same households. But it was left to feminist activists to conjoin them in public. There is perhaps an exact date for this remolding of the idea of child abuse: 17 April 1971, when Florence Rush addressed the New York Radical Feminist Conference on just this topic.

We should not be overconfident of dates. Grace Metalious's *Peyton Place,* the best-seller of 1956, is one of the most valuable social documents of the fifties. In the denouement we read that Lucas, the father of Selena, "was a drunkard, and a wife beater, and a child abuser. Now when I say child abuser I mean that in the worst way you can think of. Lucas began to abuse Selena sexually when she was fourteen, and he kept her quiet by threatening to kill her and her little brother if she went to the law" (Metalious 1956, 347). This is an extraordinary glimpse at what was happening in that most prosperous and complacent decade in American history. Yet there was no widespread self-conscious public connection between child abuse and incest until May 1977, when *Ms* magazine's lead story was "Incest: Child Abuse Begins at Home" (Weber 1977). A welter of otherwise discordant figures confirm that men sexually abuse girls in their families far more often than anyone abuses boys.

> Why, though this finding has been consistently documented in all available sources has no previous attempt been made to explain it? Why does the incest victim find so little attention or compassion in the literature, while she finds so many authorities who are willing to assert either that the incest did not happen, that it did not harm her, or that she was to blame for it? We believe that a feminist perspective must be invoked to address these questions. (Herman and Hirschman 1977, 359)

Although sexual abuse in the family became an acknowledged "social problem" only after 1975, the authors were right to speak of consistent previous documentation. Very high statistical estimates of what we now call sexual abuse of children were not new. Thus in the 1950s the second of Kinsey's famous sexual reports, the one on women (1953), found a 24% prevalence rate for girls. Landis (1956) got a 30% prevalence rate for males and 35% for females. Kinsey did not think that the early experiences need be a bad thing.

Incest is an incredibly powerful taboo. In a book titled *Incest as Child Abuse* we read that "Adult-child incest strikes at the very core of civilization" (Van der Mey and Neff 1986, 1). The traditional prohibition on incest applies to sexual intercourse. As soon as incest and child abuse came together, the concept of incest was radically extended. Fondling and touching became incest just as much as intercourse (Browning and Boatman 1977; Forward and Buck 1979, Finkelhor 1979a). Cornelia Wilbur (1984, 3), the doctor who gave us modern multiple personality, wrote that "chronic exposure to sexual displays and sexual acts during infancy and early childhood is abusive. This occurs when parents insist that a child sleep in the parental bedroom until eight or nine years of age."

The concept of child abuse also took under its wing "sibling abuse." "Evidence suggests that violence among children, especially siblings, is quite prevalent and perhaps likely to increase as more single and working parents are forced to leave small children in the care of older ones" (Wissow 1990, 195). Then sex play among children, especially with a significant difference in age, was increasingly regarded as a kind of child abuse, and hence incest. Oedipus becomes Elektra.

Liberation

Many of these disclosures were extraordinarily liberating. They made it possible for many women, and increasingly many men, to bring into the open their degrading experiences, usually at the hands of male relatives by blood, marriage, or convenience—fathers, uncles, grandfathers, cousins, step-fathers, boyfriends, companions of mother or aunt. There were also a few men who remembered forced sex with mothers and aunts. Telling the stories was cathartic. The suffering lay not just in the immediate assault and fears of the next one, but in an ongoing destruction of personality, a growing inability to trust anyone, to establish loving and confident relations with any human being. There was not only a twisting of sexual responses but also a distortion of any affectionate response. Not battered babies but battered lives.

The flip side is less attractive. Vicious disputes between divorcing parents were made to center on bad touches. Small law firms became national giants undertaking defenses of fathers who claimed they were being smeared by their wives seeking custody of the children. Even those far from the fray had a little trouble figuring out how to teach children the difference between "good" and "bad" touches. For example de Young (1988, 64) discovered that small children perceive nice touches as good, and painful ones as bad. And then there is the overall question of whether any and all sexual experiences involving children and adults inevitably harm the child. For some time the most influential scientific expert on child sexual abuse was David Finkelhor, who concluded, almost without qualification (1979b), that adults always harm children.

The various pedophile interest groups came to find spokesmen in respectable print around the same time (for example, O'Carroll (1980). Leaving them aside, it takes a detached and possibly courageous group of "experts" to insist that there are a great many different types of relationships, and that adult-child sex even within the family is by no means a blanket evil demanding automatic unconsidered state intervention in every case. Thus Li, West, and Woodhouse (1990) produced a book consisting of two separate essays, one being a survey of boys' sexual experiences, and the second a more in-depth and skeptical analysis. It is all very scholarly, but it had a great deal of difficulty finding a publisher. You just cannot say that adult-child sex might not always be so bad as is now believed.

Child abuse reminds us of a curious fact about the present state of our civilization. We are supposed to be overwhelmed by relativism. It is said that there are no more stable values. Nonsense. Try speaking out in favor of child abuse—not under the guise of man-boy love, a guise that is much spat upon in most quarters—try going the whole hog. It just does not make sense to be in favor of sexual abuse. Only monsters could be like that. But do not overemphasize the sexual here. The same happens to be true of some other items suggested by the alphabetical list with which I began Chapter 1. Take literacy. Can you imagine speaking out in favor of widespread illiteracy for the working poor? Child abuse and illiteracy are absolute (bad) values. Our society is not nearly as relativistic as is made out.

Only in our society, and only in the past thirty years, has the incest taboo increasingly stretched to any kind of sexual arousal, gratification, or implication. The extension occurred almost overnight. Why? Partly because of the link with child abuse, which was a *kind* of behavior that increasingly covered a great variety of different actions. When intercourse and exposure were included in that kind, then the subkind, family sex abuse, would be all one thing, and we have a name for that: incest. "Incest" came to mean any type of sexually oriented activity involving an adult and a child in the same family. That automatically made previously venial behavior absolutely monstrous. Next came accusations. We have just passed through a cycle in which accusations of father-daughter incest have proliferated, followed by a powerful backlash. That is another story.[8] Is there any general lesson to draw about incest and accusations?

The anthropologist Jean Comaroff (1994, 469) suggested that incest is one of the "predictable tropes of counterbeing in the late twentieth century world" (by which she seems mainly to have meant America). A more cautious observation is that the fear of incest well suits the American fear of the disintegration of old patterns of family life, and that accusations of incest confirm the fear, which then feeds on the accusations. But maybe incest is not the only idea to focus on. Many are deeply troubled by lost innocence. Incest is sexual activity forbidden to people who are related to each other. But sexual activity with children is forbidden to everyone. That is because, even after Freud, they are supposed to be innocent. That recalls the innocent Christ dying for our sins ("suffer the little children to come unto me") and the myth of Victorian Christianity about the innocence of children.

When adults in therapy are encouraged to recollect trauma of child abuse, and then to make accusations, the result is often like a Protestant conversion. Ever since Augustine, conversion experiences have been associated with confessions—the retelling of one's own past, the true past that one had been denying. All that is familiar: therapy as conversion, confession, and the restructuring of the remembrances of one's past. Then comes an almighty twist. Your confession is not to *your* sins but to your father's sins. We do not have Christ the son taking on the sins of the world. Instead the father takes on the sins that have destroyed your life. We are not concerned with Jesus, the sacrificial lamb, but with an old goat, a scapegoat, the father, the sacrificial ram.

This may sound like an excessively Christian location for the idea of child abuse. In fact child abuse activists tend to have been either feminists or dedicated Christians. The lunatic wing of the child abuse movement, uncovering innumerable cases of sadistic satanic ritual child sexual abuse, fits right into the iconography that I have just described. The patriarchal abuser is the devil incarnate, joining some extreme versions of radical Christianity and radical feminism in an unholy alliance.

Counting

When is a concept well understood? Philosophers have a trite necessary condition. If the concept applies to individuals, the criteria for applying the concept should be clear enough that one can go about answering the question "how many?" In the case of an attribute there are two kinds of "how many?" The *prevalence* of attribute A is the number of individuals in a population with A—who have been abused as children, for example. The *incidence* is the number of individuals in a given year who have A—who have been abused that year. Both may be reduced to percentages. We get amazingly discrepant prevalence and incidence rates.

Child abuse was caught up in a numbers game from the start. The lead editorial in 1962 in *The Journal of the American Medical Association* (181/1: 42) accompanied the first Denver paper. It started the ball rolling with speculations about the numbers of children killed by parental or caretaker battery. Battering was put in with the diseases, and it was (almost certainly wrongly) guessed that it caused more deaths, in 1960, than leukemia, cystic fibrosis, or muscular dystrophy.

The popular press immediately drew attention to the "tragic increase" although there were no existing data relative to which an increase in child battery could have been established. Even late in the day, with myriad data to hand, we have little sense of whether bigger numbers are more the result of more children suffering from the same maltreatment, or of better reporting of the same maltreatment, or of more events being perceived as maltreatment.

Death might provide a bench mark. Aside from infants killed within 36 hours of birth (which used to be, and in some jurisdictions still is, a separate offense) the numbers of dead children are pretty well known. There are disputes, of course, about what they died from. Abuse will be covered up as an accident, or politely disguised by a benign physician. There is the gaping question of "crib" (or "cot") death, Sudden Infant Death Syndrome. Some activists (such as Search 1988) were convinced that this is a euphemism concealing a lot of child murder. Crib death, traumatic for parents, is an entity that got turned into a "problem" (Johnson and Huffbauer 1982). It is an unusually literal case of construction, not so much social construction, as single-handed construction. Abraham Bergman (1986), a former president of the National Sudden Infant Death Syndrome Association, frankly recounted how his practice of "political medicine" turned SIDS into a well-funded problem. He called his book "a cookbook for other neglected health problems or social causes."

There are difficulties with death statistics. Yet so far as one can tell, figures for deaths caused by abuse and neglect are relatively constant.[9] That is *prima facie* evidence that the incidence of physical child abuse is not experiencing radical discontinuities or astronomical growth—and, conversely, that the vast investment in programs, agencies, and information did not change things much. The figures for abuse and neglect told a different story. The American growth rate was stunning. The first nation-wide American surveys were conducted 1967–68. The result: about 7,000 victims of abuse and neglect (Gil 1968). The estimates were based on state and local reporting, and some fairly sophisticated inferences based on polls of small populations (asking who in the group knew of at least one abused child?) In 1974 the figure was 60,000. In 1982, 1.1 million American cases of abuse and neglect were reported. In 1989, it was 2.4 million.[10] Time and again new numbers raised cries of alarm or even desperation.

The change from 7,000 (in 1967) to 1,100,000 (in 1982) is partly attributable to changing definitions. The 1967 survey was directed at physical abuse, "non-accidental physical attack or physical injury, including minimal as well as fatal injury, inflicted upon children by persons caring for them" (Gil 1968). But D. C. Gil, author of that precise definition, was after bigger game. Testifying to a Senate committee, he stated a later definition of his (Gil 1975, 20). Child abuse was characterized as anything that hinders "the optimal development of children to which they should be entitled," regardless of its cause. This much annoyed the senators, who wanted a definable, actionable, and above all nonsocial problem. Gil yearned for a "paradigmatic revolution towards non-violent societies" (Gil 1978, 31).

Gil thought that physical abuse was not a major problem. Of the 1.1 million 1982 cases, just 69,739 fell under the category of "physical abuse and neglect," that is, physical abuse *or* neglect. The figures are not precisely broken down further. But the report states that most of the 70,000 were neglected children, not physically abused ones. Which reports were correct? The National Center stated that of the figure of 2.4 million cases for 1989, 900,000 were confirmed. An opponent of "over-reporting and over-diagnosing" asserted in 1985 that conversely more than 65% of the annual incidence cases are unfounded (Besharov 1985).

Gil's original incidence rate of 7,000-odd was for physical abuse, rather strictly defined. By 1982, when there were 1.1 million cases reported, sexual abuse was firmly in the picture. Not surprisingly, incidence and prevalence rates for sexual abuse show great variation. Reputable surveys provide results for Americans that range from a prevalence rate of 6% to 60% for girls and from 3% to 30% for boys (Finkelhor 1986). Some of these disparities are readily understood. We have rates of different things. Take the seemingly simple and objective question of age when first abused. It clearly matters how old a "child" can be. Obviously both incidence and prevalence go up as age increases. In one of the most careful studies, of 2000 Canadian adults, 53.5% of females and 30.6% of males reported having had a bad sexual experience when young—but only about half in each group said that the first experience of this sort occurred when they were under 16 (Badgely 1984, 180).

There are other "objective" differences about what to count. In trying to understand the prevalence of sexual abuse, should one count indecent exposure by a stranger? From the point of view of a much-pressed social worker trying to take small ameliorating steps to help some children, incest, however understood, seems totally different from flashing. Likewise children over 16 confront us with a host of problems, but sex beginning at 16 even with a parent is relatively on the back boiler. So the social worker does not want the larger numbers.

On the other hand, if you believe that child sexual abuse is male violence to the immature, then flashing and incest at any age are all part of a continuum. Thus in a celebrated household survey in San Francisco, D. E. H. Russell (1983) found that 54% of women sampled had been victims of sexual abuse. She counted events that were remembered as happening up to age 18, and counted indecent exposure and other "non-contact" sexual abuse. But her purpose was quite straightforward, to document the extent of male terrorization. Child sexual abuse is an instance, along with sexual harassment and rape, of the male violence that at present threatens the "well-being and survival of the entire population of the United States" (Russell 1984, 289). It does not help so much to say that if you count different things you will get different answers, for what you count depends upon your theory about you are counting.

Export

Christians have exported their sexual mores around the globe, preaching monogamy and creating modest clothing, such as the muumuu for Hawaiians. One of the most striking epiphenomena of child abuse is its missionary element, its desire to carry the bad tidings to other nations.

It is a truism that a great deal of behavior that we hold intrinsically loathsome and terribly harmful to children is merely venial or even encouraged in other cultures. That was well known before child sexual abuse was on the scene as a confirmed "social problem." In the second of his famous reports, Kinsey (1953, 121) remarked *a priori* that "it is difficult to understand why a child, except for its cultural conditioning, should be disturbed at having its genitalia touched, or disturbed at seeing the genitalia of other persons, or disturbed at even more specific sexual contacts." And already by the same date one could, *a posteriori,* find in a single volume derived from the Human Relation Area Files of Yale University enough sexual practices and codes of ethics to dazzle the most jaded enthnographic voyeur (Ford and Beach 1952). A later contributor to the ongoing collection of this sort of information observed that "the inherited aspects of human sex seem to be nearly formless, only by enculturation does sex assume form and meaning" (Davenport 1976, 161). This is as true of adult-child sex as anything else.

Such reflections have not hindered the movement against child abuse from exporting its concerns. It has been an article of faith among many American activists that child abuse has been occurring in most cultures at most times. Child abuse, as a diagnostic and political concept, has chiefly been a phenomenon of the English-speaking world, with the United States as almost the only source of conceptual innovation. Despite its regional character, the movement organized itself as international. The first professional journal dedicated to child abuse was founded in 1976: *Child Abuse and Neglect, The International Journal.* In parallel there was established the International Society for Child Abuse and Neglect. The headquarters was to be Geneva, where else for an International Society? The president was a Swiss pediatrician who had contributed almost nothing to public knowledge of child abuse.

Around-the-world meetings were set up. In 1986, Australia. In 1988, Brazil. That enabled the president to announce glowingly that concern with child abuse was moving into the southern hemisphere (Ferrier 1986). Similar enthusiasm accompanied the First European Conference on Child Abuse, Greece, 1987, advertised as having the special bonus that it would involve delegates from East as well as West Europe. It is striking how often studies of child abuse have employed the metaphor of new territory—"the last frontier in child abuse" (Sgroi 1975), or "the hinterland of child abuse" (Meadow 1984). The International Society, headquartered in Geneva but driven from the United States, had a clear vision of manifest destiny.

It was not so easy to go international. Literature surveys, always conducted in English, agreed that child abuse was not in fact widely seen as a threat. "In Poland, where child abuse is not yet defined as a separate problem . . ." (Kamerman 1975, 36). Two of the leading figures in the movement noted that "Our review of the literature on child abuse in other countries found considerably less concern for estimating the incidence of abuse than there has been in the United States" (Gelles and Cornell 16). The paragraph continues by saying that the objective methods of estimating incidence that had been developed in America had only once been used outside the United States. There was some recognition that "as norms and attitudes vary, so do the research efforts, data collection mechanisms and knowledge generated about family violence."

Third-world studies were of two sorts. One was ethnographic (*we* study *them*): "Child abuse and neglect—rare but perhaps increasing phenomena among the Samia of Kenya" (Fraser and Kilbride 1980). At the other extreme is the serious attempt by someone outside the movement, but inside another country, to say we have problems about children, but not your problems. The Head of the Department of Nutrition and Metabolic Diseases in the Calcutta School of Tropical Medicine, described as "the only third-world participant" at the 1978 International Conference on Child Abuse and Neglect, had been able to find only four published cases of battered children in India. Instead he talked about food distribution. There is no shortage of food in the world, but agricultural technology has terribly outstripped our collective ability to feed all of us. His paper "characterizes two hitherto unrecognized syndromes of prolonged and protein-energy malnutrition, suggesting the term 'nutritionally battered child' for the victim of either syndrome" (Bhattacharya 1979). This physician did not mean anything like the American battered child, that is, a child harmed by the assault or even neglect of its immediate caretakers. He was talking about the 40% of the (then) 115 million Indian children living below the poverty line (family income less than 72 rupees, then about $6 a month).

The drive for expansion sometimes backfired. Foreigners started saying that child abuse is an American problem. What else would you expect from such violent people? Where American influence was strong,

Child Abuse and Neglect Reporting Centers were set up on the American pattern, but not with the same results. In South Korea, despite much publicity, a total of twenty cases were elicited for 1988 (Chun 1989).

There was nevertheless a calm confidence that the rest of the world would see the American light. At the 1978 international congress, Kempe read into the record a universal sequence of events for any society whatsoever. First, there is denial. Then lurid abuse will be admitted, as with battered babies. After that, physical abuse of all sorts, and then "failure to thrive syndrome." Then people will move on to a fifth stage, of considering emotional abuse and neglect. Finally there is the sixth stage of concern, in which society tries to assure each child that it is truly wanted. Even in America we have not got there yet, said Kempe, but we have (1978) come to acknowledge that sex abuse and incest are as common as physical abuse. Every country must go through this recognition at its own pace (Kempe et al. 1980, xvi–xvii).[11]

Objectivity

Those human kinds—"child abuse," "abused child," "child abuser"—have been molded and revised in the United States and then exported. There is one astounding exception to the rule, an entirely home-grown British contribution. *Lancet* carried a general discussion of a well-known technique, known as anal dilation, used for recognizing anal sex practices or assault in adults. Hobbs and Wynne (1986) reported that it could also be used with small children: "Buggery in young children, including infants and toddlers, is a serious, common and under-reported type of abuse." The method consists in an abnormality in the anus observed upon separating the cheeks of the buttocks. The article plainly stated that "the specific forensic examination must take place in the context of the whole child examination which in turn forms part of the assessment of the family as a whole."

Then some consultant pediatricians transferred anal dilation from standard forensic practice to clinical diagnosis. They worked in hospitals in the industrial city of Leeds. Hospitals in that region had been a British leader in diagnosing physical abuse of children, gladly acknowledging debts to American precedent and advice. In 1981 Leeds registered 5 cases of sexual abuse. In 1986, 237 of the cases referred to Leeds hospitals were confirmed as cases of sexual abuse, and the number doubled the next year, the time of publicity of what in the U.K. came to be called the "Cleveland affair."

The work in Leeds would not have been much noticed by the general public were it not for the dedicated work of two consultant pediatricians in a group of pediatric hospitals in the Northeast of England. Between February and July 1987 they diagnosed sexual abuse in 121 children, 67 of whom became wards of court. A further 27 were temporarily removed from their families by "place of safety" orders, direct interventions by the social services against which the family has no immediate legal recourse. More than half the children involved were under six years of age.

The public reaction was completely the opposite of what one would have expected in the United States at that time. There, almost all accusations of child abuse gained widespread credence and were usually supported in the popular press—even when the conviction rate for the weirder stories was not high. In Britain the tabloid newspapers rounded on the pediatricians involved. Their Member of Parliament defended the parents, accusing the authorities of engaging in a crazed witchhunt—Salem itself was invoked by a "conservative" Labour Party M.P., drawing his support from traditional trade unions (Bell 1988). "Radical" Labour city councillors defended the pediatricians, citing "international" statistics to prove that there must be just the sort of incidence of sexual abuse that the doctors were discovering (Campbell 1987, 1988). This was, incidentally, a nice example of the export market, as all the international statistics were from North America. Incest Crisis Line, founded in 1987 as one of the very first "hot lines" for victims in Britain, suddenly found itself taking 1,500 calls a week from incest victims.[12] A commission of inquiry was established. It presented an extremely cautious but fairly precise definition of child sexual abuse (with no mention of incest:

> Sexual abuse is defined as a the involvement of dependent, developmentally immature children and adolescents in sexual activities that they do not fully comprehend and to which they are unable to give informed consent or that violate the social taboos of family roles. In other words it is the use of children by adults for sexual gratification. Dr Cameron described it as inappropriate behaviour which involved: "the child being exploited by the adult either for direct physical gratification of sexual needs or for vicarious gratification." (Butler-Sloss 1988, 4)

Of the 67 children who did become wards of court in the Cleveland affair, proceedings were dismissed for 27. By July 1988, 98 of the 121 children were back home. It was inferred by the tabloid press that the pediatricians had been wrong, although in fact the report supported almost all the diagnoses themselves, with two exceptions. Most of the children placed in care had already been drawn to the attention of the hospitals and a good many would have been in the hands of the social agencies solely for suspected physical abuse.

What was wrong was that the separations were made immediately after diagnosis. No one explained to the children or families what was going on. There was an overwhelming air of arbitrary decision, even if the doctors and social workers truly cared about one aspect of the children. These authorities were unable to grapple with larger social realities—like what to do with this sudden influx of children who had been placed in care. And they seemed almost indifferent to the emotional pain for children and families who were separated.

There have been many analyses of this catastrophe, ranging from the witch-hunt theory to complete support for the pediatricians. The middle-of-the-road view is that the cases were botched, that anal dilation is at most an indicator of possible trouble, that it should never be used in connection with children except as a sign that a family should be scrutinized more carefully. And, probably, most of the children were being abused, but peremptory separation with inadequate back-up services was not the way to deal with the problem.

The case illustrates how the concept of child abuse craves objectivity. In the natural sciences we think there are objective criteria for telling whether or not an item is of a certain kind. That was why battered child syndrome had been so convincing: it could cite X-rays, proof of science at work. Child-abuse workers are frustrated by a lack of agreed criteria, of "scientific" proof. When a child catches a venereal disease or semen is found in its body or even clothing, that is objective all right, but too rare for general use. Even though an involved person—be it neighbor, relative, teacher, or specifically empowered agent such as pediatrician, police officer, social worker—is sure that a child is being abused, we often lack agreed public criteria for demonstrating the abuse.

Anal dilation was the magic solution for a class of crimes against children. A simple clinical observation made in moments was decisive. Objectivity had finally entered the domain of child sexual abuse. The consequence of objectivity was catastrophe.

The place-of-safety orders followed all too automatically from the first precept of 1962, quoted above, that when a child is abused (in 1962, battered) it is better to separate it from the caretakers, no matter what else. The child is bewildered, but better to make the sharp break then and there. There is in addition a further therapeutic theory, that it is essential for the child to own up, to express, the facts of abuse and its own emotional reactions. Coupled with this is the experience that such confession can be done only when the child is able to be away from the abuser for some period of time. The radical shortage of welfare workers and therapists to help heal the wounds and work through the mess was recognized by the Cleveland pediatricians as a bad thing, but not sufficient ground for inaction. Leaving the child at home after being seen by doctors would, they thought, only reinforce the idea that the doctors were colluding with parents. If doctors and social workers did nothing, the child would see that it was helpless, with no one to turn to. Hence a place-of-safety order.

The reasoning seemed impeccable, but the results were disastrous. There is conclusive evidence that the pediatricians became obsessed with their mission, resentful of colleagues, and increasingly indifferent to parents. They became so fixated on one set of obligations that they quite literally forgot the original message in *Lancet*, quoted above, "the whole child examination which in turn forms part of the assessment of the family as a whole." But at a greater distance we can make an observation about objectivity.

A standard reaction to this case is that anal dilation is "not reliable" as a method of diagnosis. That may not be the right reaction. The basic trouble with some classifications of people is that objective identification of instances of the kind of person in question misses what is important about the kind, and deludes us into thinking that a straight and simple road is to hand.

Figures of Speech

Often the ways in which a new kind is selected as relevant involves what, in other contexts, we call figures of speech. Some metaphors do not catch on. Thus the metaphor of the nutritionally battered child, proposed by the Indian pediatrician to describe malnourished children in the subcontinent and elsewhere, fell by the wayside. This metaphor was not fueled by the deep passions of innocence, incest, and the collapse of nuclear family; it was just millions of hungry children of no significance.

Child abuse served as a cutting-edge metaphor closer to its home. With its ramifications in sex, beating, and emotions, it does not pick out one kind of behavior. It is a kind whose power is to collect many different kinds, often by metaphor. This power can be put to use by many an interested party. At the time of the 1990 Gulf War the spokesman for the Kuwaiti government in exile stated for television viewers of the West that his country was a small, abused, and molested child.[13] A man in Charleston, West Virginia, unhappy with the way his town was planting trees in the sidewalk, growled, "We have child abuse—this is tree abuse!" and founded a Society for the Prevention of Cruelty to Trees.[14]

That is a bad joke. Missing children are not. But they too represent a cause that uses child abuse as a metaphor. The advertisements for "missing children" that in the early 1980s plastered American cereal packages, chocolate bars, milk cartons, and other family artifacts, not to mention direct mail and posters at laundromats and bus stations, were presented as trying to save victims of child abuse. In fact a large proportion of the advertised missing children arose from custody disputes. The child was missing because the other parent had taken it and would not reveal its whereabouts. But the innocent public thought that all those children had been abducted by child abusers.

Events of consequence were tied to child abuse. Anyone with any experience of neonatal pediatrics knows the horrible moral problems of providing care for seriously damaged or impaired babies. Many can be kept alive, although at great cost, and there are not enough resources to keep alive all those that could, in principle, survive. So hard decisions are made daily. Ignoring this reality, the law-makers metaphorically declared that suspending or limiting care for the newborn was child abuse. On 3 February 1984 the U.S. House of Representatives voted 396 to 4 in favor of amending the definition of child abuse to include any denial of care to newborn infants with life-threatening handicaps.

"Child abuse" is a potent metaphor because it has the property of instantly concealing its use as metaphor. Once something is labeled child abuse, you are not supposed to say, wait a minute, that is stretching things. Which labels stick depends less on their intrinsic merits than on the network of interested parties that wish to attach these labels. Thus drugs are the American bogeyman, not alcohol. So we have had prosecutions since 1986 for "foetal abuse" under child-abuse laws—a pregnant women is doing drugs. But of course that is tricky, since pro-abortion parties see they are being got at if, in this instance, a foetus becomes a child in the eyes of the law. So in 1989 one jurisdiction began arresting the mother at birth, before the umbilical cord had been cut, but after the child had been delivered, on the grounds that she was in the act of giving drugs to a minor (an offense in itself) and also abusing a child. On the other hand, foetal alcohol syndrome was not being put under the heading of child abuse. It is a real scourge among the aboriginal populations of North America. Every bar in America now warns pregnant women against consuming alcohol, but those who do are not charged with child abuse. Only drugs weigh in with the moral clout to merit the use of child-abuse legislation.

Other figures of speech come into play as a new kind, or a newly molded kind, is selected as relevant. Metonymy is defined as "the use of the name of one thing for that of another, of which it is a part, and which is thereby suggested by it." It is a trope in which a container is used as the name for what is contained, an association is used as the name for what is associated with it, or the part is used as a sign for the whole. It can be powerful because what is true of the part is then taken to apply to the whole. An example is the use of the word "incest" to name all child abuse within the family, with any sexual connotation whatsoever. Every touch or exhibition then acquires the horror of incest. The name of one thing, "incest" (understood in its literal pre-1970 sense), is used as the name of another thing of which it is a part (acts within the family that have some sexual connotation). This is one of the ways in which a kind can be molded by metonymy. Not constructed, but molded.

Drawing the Line

Not every sexual harm done children was allowed to count as child abuse. Sex rings involving children did count as child abuse. By extension, or perhaps simply out of a desire to use the child abuse movement and child abuse law to control pornography, child pornography became regarded as a particularly vicious type of child abuse, authorizing the American authorities (including the Postals Inspectors) to mount sting operations on foreign pornographers, even offshore.[15] That may be because child pornography was seen as part of the crisis in the family which was the leitmotif of American social self-knowledge in the 1980s.

Child prostitution, in contrast, did not figure on the child abuse agenda. Perhaps this is partly because it is outside the family. Suppose that incest was central to the molding of the idea of child abuse. Suppose that this was because incest provided a way for some feminist activists to attack patriarchal authority, and, on the larger scene, summoned up worries about the breakdown of the nuclear family. Then any child sex within the family will count as sexual abuse, no matter how the metaphor is stretched. Think of today's extended family in a generous way, as including day care, schools, boarding schools, choirs, boy scout troops and the like. Child sex within the extended family is regularly counted as the worst sort of child abuse. But hurtful sex with children and adolescents outside the extended family is not part of the child-abuse prototype. So child prostitution did not get included, in any serious way, within the bounds of child abuse.

The one exception was Badgely (1984), dealing with Canadian data. Strongly criticizing the "ambiguity, myths and hypocrisy" (p. 947) with which juvenile and child prostitution was cloaked in that country, it catalogued without comment numerous kinds of exploitation, violence, as well as health and drug risks to which child prostitutes are subjected (pp. 1007–1028). About two-thirds of Canadian prostitutes under 16 were female (p. 969). Most of the girls worked for pimps, while the boys were mostly freelance. A girl working 40 weeks earned on average $53,000, of which she kept one seventh (pp. 1057–1072). This is more than the pimps (average age 24.7) with their meagre education could earn doing anything else. The earliest age at which a child had started work was eight.

Most of the children reported severe violence from both pimps and tricks. They received essentially no help at all from social welfare agencies, and yet are probably the children most at risk from virtually every form of abuse current in our society. Some of the report's strongest recommendations concerned child and juvenile prostitution (91–98). But I should also mention a minority, unpublished opinion from some of the people who did the street work for the Report. Many, possibly a majority, of the children regard themselves as successfully exploiting nerdish men, a risky business indeed, but one which do-gooders will only make more dangerous, and which would cost some of them their livelihoods.

The Canadian study was, at the time, the only reasonably sound statistical survey. Contrary to what was uniformly announced by the child-abuse movement on the rare occasions when it turned its mind to child prostitutes and runaways, it was *not* the case that children in the sample had been more commonly subjected to family child abuse than the rest of the Canadian population.

OLD WORLDS

"Worlds differ," wrote Goodman, "in the relevant kinds they comprise." My sketch surely shows that we have evolved a new and relevant kind, namely child abuse. What has this done to our world? Let us start with the past. Every generation writes history afresh. It is to be expected that a new conceptual scheme, like that of child abuse, will be used in rewriting some history. To what extent can we use the idea of child abuse in redescribing a past that had very little idea of our present repertoire of vices?

Let us begin with people in the not so distant past. Alexander Mackenzie was one of the great European explorers of North America, after whom one its three greatest rivers is named, and who, following another river part of the way, led the first European party to cross the continent (1793). "Mackenzie was undoubtedly daring but he was also a racist and, like a number of his confrères, a child molester, marrying at the age of 48 a perky fourteen-year-old Scottish lass."[16] And other remarks about Mackenzie as sex abuser. The sentence is handy because "racist" was also no word of the eighteenth century. How do Mackenzie as racist and Mackenzie as child-molester compare?

There is no doubt that Mackenzie, like almost all his "confrères," felt that white people are superior to natives, that their interests were more important than those of natives, and that only in exceptional cases could any red man be placed in authority over any white person. It was all right for white explorers to sleep with Indian women, but repugnant to think of an Indian man marrying a European. Mackenzie, in short, was a racist. Had it been possible to explain the term to him, shorn of its present overwhelmingly negative connotations, he would hardly have wanted to deny the epithet.

But child molester? We have no evidence of cruelty, of his forcing himself on the girl or even exploiting her so much differently than an older woman. We may now find it repugnant that the British age of consent was not raised from 12 to 13 until the end of the nineteenth century, but that does not make men who married fourteen-year-olds a full hundred years earlier than that into molesters.

On the other hand, was Lewis Carroll (Charles Dodgson, the author of *Alice in Wonderland*) a pedophile? He certainly made a great many photographs of naked girls, and his diary entries, where he used the French "sans habilement," indicate that in his books this was not run of the mill photography. If, as Anson (1980) has claimed, he had a great collection of pornographic materials produced by salacious commercial pornographers, then Lewis Carroll does not seem so different from today's stereotype of the pedophile. But suppose that is only a libel against Lewis Carroll. He was just a mathematics teacher at Oxford University, whose fantasy life was populated by small girls, a fantasy life which he fulfilled by taking masses of photographs of naked innocence. If so, he can still be called a pedophile in the literal sense of the words: "an adult who is sexually attracted to a child or children" *(American Heritage Dictionary).*

There is of course a difference between calling Lewis Carroll a pedophile (whether or not he fondled anyone improperly) and calling Mackenzie a racist. To call someone a racist is to imply at least a lack of respect for the dignity of individuals of other races. To call a man of power, who is in constant contact with another race, a racist, is to imply that he harmed the weaker people he encountered. We may be content to think exactly that of Alexander Mackenzie as he traversed North America from East to West and from center to North. Are we content to call Dodgson a pedophile? Pedophile does not mean, for most people, what the *Dictionary* says it means. It means bad, dangerous, monstrous. When a block of public housing learns that a released pedophile is to be moved into an empty flat, a riot ensues. But if Dodgson just had fantasies, and filmed little girls without even implying harm, many will feel that he hurt no one, and resist the pejorative connotations of pedophilia. On that view it is especially sad that Carroll's name was taken over by pedophile organizations and publications that he would have found grossly offensive. Sadder still that a New York pedophile newsletter, which tried to keep on the right side of the law, was named *Wonderland,* and in the end it was completely taken over by the United States Postals Inspectors for entrapment purposes—as part of a master plan they code-named Project Looking-Glass, confirming their own adage that porn breeds porn (Tate 1990, ch. 8).

Now turn from individuals to communities. I have urged a contrast between the Victorian ideas of cruelty to children and our own child abuse, but many writers strike out in the opposite direction, seeing the ancient Greek practice of infanticide as especially horrible child abuse (Radbill 1968). "Cruel" fits, for sure. How could one deny that the parents or governors who exposed the infants were cruel in so doing? But child abuse? Some have no scruples. Donovan (1991) argued that Freud made a big mistake about the Oedipus legend. It is not a story of adult mother-son incest, but of rank child abuse, what with Jocasta maiming Oedipus and then handing him over to a shepherd to be killed.

The idea of child abuse is too caught up in a web of present-day causal and moral speculation for it to make good sense to use it in indiscriminate descriptions of the distant past. Personally I find Carthage utterly loathsome; its slaughter of its own children is the work not of men but monsters (my mind may inform me that *Salammbô* tells more about Flaubert and nineteenth-century French Orientalism than about the end of an ancient city, but a great novelist overrides the mind). Did the Carthaginians systematically practice child abuse? That would make sense only within a much larger historical framework.

We do have such a framework, or rather two of them, classic statements of opposed opinion about children. Philippe Ariès (1962) had a vision of earlier centuries of childhood as providing a freer, franker, less sexually cluttered life for humans before and shortly after puberty. There was not much of an idea of the child, and still less of abusing. Humans in that age group and were not harmed, were not conceptually capable of being harmed, in ways that we now harm children. The psychohistorian Lloyd deMause (1974) held that to be rubbish. The history of at least Western civilization is the history of child abuse. Things get worse and worse the further back in time we go. Ariès used the constant public playing with the genitals of infant and child Louis XIII as evidence of an absence of oppressive conceptualization. (One Héroard, a court doctor of Louis's father, provided ample description of the fun). DeMause took the story as evidence of rampant child sexual abuse. Needless to say Greek pederasty fills his bill too, along with infanticide and the so-called Children's Crusade.

DeMause is nothing if not a synoptic thinker. Ethnologists worry about incest, its causes and its taboos, but not deMause, who holds that there is nothing to explain. Incest is a human universal, rampant among all peoples at all times; what would need explaining is absence of incest (deMause 1988, 274). So grand a theme requires taking "incest" in its most general possible post-1977 sense. It includes for example the widespread Victorian practice of administering frequent enemas to children. (Anal penetration administered by a family member = incest.) We are blessed that deMause seems not to have tumbled on the fact that the aforementioned Louis XIII, grown-up, subjected himself to 214 major enemas in a year (Raynaud 1862, 143). His son, the Sun-King, who with most of his court went in for the same noncures, seems to have enjoyed a burlesque whose jokes are largely based on enemas, namely *Le Malade imaginaire*.

The theses of deMause and of Ariès do matter. The one makes a claim about the nature of our species—the child-abusing species. The other makes a subtler claim about the nature of our changing conception of ourselves, of what it is to be a person. When we look back on famous men who are long dead, it matters very little to anyone except biographers whether Alexander Mackenzie was a child molester, whether Lewis Carroll was not a pedophile but a pederast, or whether the kings of France enjoyed their enemas. But there is one "personal" accusation that does matter, because the person looms so large in our present vision of ourselves. I mean Freud, and Jeffrey Masson's (1984) famous allegations about him. Freud first attributed hysteric symptoms to child or infantile incest and paternal assault. Later he decided that fantasized rather than real seductions and molestations were more often causing neurosis. But, asserted Masson, Freud and his peers had ample evidence of sexual abuse and (guiltily) refused to take cognizance of it. Was that not a moral failing, a deliberate, as Masson puts it, "assault on truth"?

The standard "defense" of Freud is the wet observation that Freud did not really abandon the seduction theory; look, in 1931 he was saying "Actual seduction, too, is common enough" (Hanly 1987). Exactly this defense is repeated by Leonard Shengold (1989, 33–38, 160, 169) in his Freudian study of the effects of childhood abuse and deprivation, which, curiously enough, is a work that is parasitic on the child-abuse movement.

Such defenses completely miss one of the sources of Freud-hatred: he never took seriously the idea of widespread paternal assault on children, assault which demands direct state intervention. This is readily added to the groundswell of intense feminist criticism of Freud. But it is a distinct allegation, and can to some extent be discussed independently of Freud's male chauvinism.

I suggest that Freud's failure, if there was one, was less a matter of dishonesty than lack of a larger organizing category. Once again I make a distinction between cruelty and child abuse. In their lascivious search for perversions, nineteenth-century doctors did turn their thoughts to child molesters, pederasts, and pedophiles. But these individuals were not linked with cruelty-to-children. The treatment of such individuals was kept to the clinics, while the control of child beaters (cruelty-to-children) was arranged in the courts.

Perverts of various sorts were very much "kinds of people" for nineteenth-century doctors. But cruelty-to-children was not a nineteenth-century scientific kind. In 1890 there was no larger "kind" into which the incestuous caretakers of Freud's patients could be tidily fitted. There was also nothing like the London anti-cruelty movement in Vienna; witness the reactions to yet another pair of much-reported dead children in the Vienna of 1899, described by Larry Wolff (1988). The two children, one rich, one poor, had been murdered, brutalized, neglected. Sexual abuse was not evidenced in court (which is not to say it was not hushed up).

Wolff takes the trials to be relevant to Freud's seduction theory. Wolff may be guilty of a common anachronism. In New York, in the 1980s, sexual abuse and physical abuse were the "same kind" of thing, so "of course" Freud would or ought to have seen how the trials bore on the seduction theory. But in Vienna in the 1890s the two kinds of abuse were not deemed to be alike. If anything, incest was taken to be confused or deviant affection, while cruelty was its complete opposite. Soon after founding the satirical broadsheet *Die Fackel,* Karl Kraus, the radical critic of the Austrian world, used the cases of the two dead children for his attacks on Viennese hypocrisy, but always it was on the score of physical cruelty and indifference to children, not sex.

I should not be astonished if the Ringstrasse in Vienna encircled one big child sex ring that a great many people knew about. But in those days, if that vice had come to light, it would have been kept quite separate from cruelty to children, rather as today the child-abuse movement keeps child prostitution in a different box from intrafamilial sex. Wolff has been criticized for his high-flying speculation that Mahler's *Kindertotenlieder* of 1901–1904 was prompted by the murdered children. Whether Wolff is right or wrong about that, the speculation is logically coherent. Children were murdered. Mahler wrote a magnificent song cycle about the death of children. Murder and death are of the same kind. But the murder of the children was not conceived of as child abuse—in

1900. There was plenty of child abuse about in 1900. But that classification, *our* classification, was not yet in place.

NEW WORLDS

The kind "child abuse" has created a world of difference. Children are subjected to education about it, by way of videos, from the earliest years of schooling. Television and movies have a steady diet of it. There are support and confessional groups for abusers, modeled on the lines of Alcoholics Anonymous. Abuse has been firmly grasped by co-dependency movements. By 1985 there were cities—Portland, Oregon, for example—in which anti-abuse activists had been so successful that men were advised never to touch a child in public; if a child not in the family is hurt, be sure there is a friendly witness before helping it in any physical way.

Looping Effects

Erving Goffman (1963) speculated that the classification of deviants by social scientists might reinforce or even engender deviant behavior. Child abuse certainly invites speculation along the lines of labeling theory, that is, people come to see themselves as abusers by being so labeled. More interestingly, Schultz (1982, 29) has argued that symptoms of being abused may be iatrogenic, that is, induced by the helping professionals who work with a case of child abuse. "The very labeling and intervention in child/adolescent/adult sexual interaction may themselves be victimogenic or traumatogenic." That means that the child experiences trauma or experiences herself as victim only after her life is treated as a case. "Labeling the child a sex victim, or assuming a symptom complex may have self-fulfilling potential" (Schulz and Jones 1983). That would be an instance of the looping and feedback effect of the evolving kind, child abuse. C. K. Li (Li, West, and Woodhouse 1990, 177) notes how more research generates more experts generates more cases generates more research . . . "It is apparent that a positive feedforward cycle has been operating here." (Positive feedforward cycle is what I call looping.) Li drew attention to the belief that more is always truer. Discussing two papers on how to conduct child-abuse surveys (Wyatt and Peters 1986a, 1986b), he wrote: "Although not explicitly stated, the aim is to produce more 'accurate,' i.e. higher prevalence rates—the assumption is that, since child abuse is ubiquitous, only higher rates are truly indicative of the magnitude of the problem. The possibility that a survey interview may become a process of persuasion is simply ignored . . ." (p. 179).

Self-Knowledge

I wish to conclude not with looping effects, however, but rather with the far more difficult notion of self-knowledge. We may have trouble retroactively applying child abuse, as a kind of behavior, to people or periods in the past. But common sense is no bad guide. It cautions us against the heady brew of Ariès and deMause alike. It tells us that Alexander Mackenzie was no child molester but that Lewis Carroll was a (quite probably harmless) pedophile. Common sense is not, however, well-honed to applying a new concept, a new kind, to our own past.

One of the most striking consequences of the post-1975 uncloseting of family sexual abuse is that many women and quite a few men now see themselves as having been sexually abused. Many feel a great relief: finally they are able to talk about their experiences. Some resent being forced to recall what they had repressed. But there is also the phenomenon of retrospectively seeing events as abusive which were not directly and consciously experienced as such at the time. It is only dogma, which degrades the complexity of human consciousness, to say that they always were known to be abusive, but the knowledge was covered up out of fear or indoctrination. Sometimes that is the right description. Tens of thousands of women know perfectly well what was done to them. But we are also witnessing or have just witnessed a radical re-evaluation of childhood experience, a reclassification, and in a way a re-experiencing of it.

There are plainly two extreme options, each attractive to one ideology or another. One says that if our consciousness is now raised so that we see an event as abusive, then that event always was abusive, even if no one intended it that way or experienced it that way when it occurred. That is what consciousness raising is all about! The other option resists this, and says the events were not evil in their time, though it would be wrong to repeat acts like that now.

Does it matter who is right? It matters to prevalence statistics in a straightforward but philosophically uninteresting way. For if we do a survey to find out how many children were abused, we are asking adults about their past as seen from here. D. E. H. Russell took the former, consciousness-raising, option, and she got the highest prevalence figures of anyone (Russell 1983). Entirely rightly, from her philosophical perspective, she took particular care to use only female interviewers who were especially "sensitized" to intra-familial child abuse. These are precisely the interviewers who will enable a person to see past events in a certain way. Using these assistants Russell was, on her account, the "first to document thoroughly and rigorously the extent of sexual exploitation in a major U.S. city." She concluded that "every second female in San Francisco has been sexually abused" (Russell 1984).

Numbers matter to all sorts of questions of policy. No doubt numbers like Russell's contribute to Li's "positive feedforward effect." But more pressing than numbers are the ways in which individuals must now deal with as difficult a question of personal reality as has ever afflicted a large group of people. What happens to the woman who now comes to see herself as having been sexually abused? I am not now referring to the person who has merely kept an awful private secret, who now may feel liberated by being able to talk about it, or oppressed by having it brought to surface consciousness again. I am referring to entering a new world, a world in which one was formed in ways one had not known. Consciousness is not raised but changed. Someone now sees herself as abused as a child, because she has a new concept in terms of which to understand herself.[17] This is perhaps the strongest and most challenging application of Goodman's dictum, that worlds are constituted by kinds. Child abuse is a new kind that has changed the past of many people, and so changed their very sense of who they are and how they have come to be.

Chapter Six

WEAPONS RESEARCH

One argument in the "science wars" goes like this: the velocity of light is a fundamental constant of nature. It is about 186,282 miles per second. We know the actual value to a very high level of exactness. This number is completely independent of any social circumstances whatsoever. All the major contributors who have helped to determine this number to several places of decimals, are dead white males. But essentially the same measurements would have been obtained if the investigators had been women or Polynesians.

That is Internet talk. I have run together, cleaned up, and shortened several versions that one comes across. The distasteful reference to women and Polynesians is a muddle. Some scientists who dislike constructionism think it has something to do with multiculturalism and the argument that college students studying literature should read more than the classical canon of European and American authors—and that history students should learn more than the exploits of the heroes of a few chosen nations. Multiculturalism has nothing to do with the argument. The point of the velocity-of-light example is that anyone who seriously asks, "what is the velocity of light?"—anyone who industriously investigates the question—will get the answer that has already been obtained.

The claim seems to be that *if* some people ask this question, and work hard at answering it, they will get the standard answer. I wish to address a quite different issue, but we do have to notice that the claim is false. The thought (which did occur to Galileo) that light has a finite velocity, is itself a remarkable one. Even a group of investigators who had that thought, and had to invent their experiments and make their equipment, would be very unlikely to get any answer at all. Measuring the velocity of light is not a piece of cake. So let us try to sharpen the claim: if some people ask the question, and work hard at answering it, and get an answer, they will determine that the velocity of light is about 186,000 miles per second.

This claim is also false. The first man to establish the finite velocity of light was the Danish astronomer Ole Römer (1644–1710). He was a gifted observer, brilliant instrument maker, and astute theoretician. He was also extremely well connected, and had worked with the best astronomers in Paris, not to mention tutoring the son of Louis XIV. He was much preoccupied with the motions of the moons of Jupiter, and especially with their eclipses. The moon named Io was of particular interest. The elapsed time between the observation of its successive eclipses, when the earth is moving away from Jupiter, is different from the elapsed time when the earth is moving towards Jupiter. But the relative speed of the two planets does not completely account for the discrepancy. Using inferences about the velocity of Earth's approach and departure from Jupiter, he invoked another factor to explain the time differentials—the velocity of light. He calculated the velocity to about 140,000 miles per second, which every astronomer alive today respects as a brilliant estimate. But Römer, one of the most talented and scrupulous observers of all time, did not get 186,000 miles per second. Hence the statement in the Internet rhetoric is false.

What claim is true? That if other people were to use our equipment, with our assumptions, and had acquired all the tacit knowledge needed to use our equipment, they would get our answer? Not even that. For if they got another answer, we would surely rule that they had made a mistake. So is the claim that if they used our techniques and made no mistakes, they would get our answers? We are close to an empty platitude, a tautology. Nevertheless, let us respect the instinct behind the Internet claim, even if does poorly when subjected to serious challenge. Suppose something like this is correct: if a group asks about the velocity of light, and works hard at it, has plenty of material, intellectual, and cultural support, and does everything pretty much as has in fact been done, then it should get something like present-day measures of that number.

If a group asks the question. That is a big "if." Such a question makes sense only within a remarkably specific context. It is clear that no human beings need ever have thought of the question, or believed that it could have an answer. The senses teach, if the question is put, that the propagation of light is instantaneous. Questions about the velocity of light need never have arisen. What suffices to make a question legitimate? That is the topic of Nicholas Jardine's (1991) book, which is subtitled *On the Reality of Questions in the Sciences.* He shows, by means of compelling historical examples, how questions that make sense in one scientific framework are unintelligible in another. The frameworks in which we ask questions arise out of a historical process, and seem to be contingent in the sense discussed in Chapter 3.

We might say this. Once a question does make sense, its answer is determinate. That is perhaps the force of the confused Internet claim. The answers to questions are the content of a science. Given the questions, the content may be fixed. But what questions will make sense is not predetermined. This chapter develops an idea parallel to Jardine's. Instead of speaking of questions and answers, it speaks of the content of the science, and the framework or form of the science within which questions can be posed and answered. In my questionnaire at the end of Chapter 3, I scored a feeble 2 out of 5 on the side of the contingency of scientific results. That is because I am not a contingentist about the content of science, once the questions are intelligible and are asked. But I am inclined to contingentism about the questions themselves, about the very form of a science.

This chapter explains why. It was written for a special issue of the *Canadian Journal of Philosophy,* sent to press in 1986, commissioned two years earlier, and published in 1987. The topic for this special issue was the ethics of warfare, especially nuclear warfare. I had nothing to say about that, except well-meaning commonplaces. The year 1984 was the heyday of the American Strategic Defense Initiative, popularly known as Star Wars. So I turned my attention to what weapons research might be doing to science, not as affecting the content of science, once questions were asked, but as affecting the questions to be asked. Other thinkers have addressed similar questions, for example Evelyn Fox Keller, both abstractly (in her 1992), and concretely in her examination of the way in which genetic research has been formed by the image of codes that control the development of each living thing.

As with Chapter 5, I have not updated this essay. The point of using it here is not to make a timely observation about military research. And I certainly am not arguing for the stupid idea that if peaceniks had asked the same questions as investigators funded by the military, they would have got different answers. I expect they would have got the same answers. Indeed, many moderate peaceniks have been supported by grants that can be traced back to the Department of Defense. The point of using the discussion here is to take military questioning of nature as one example to illustrate the difference between the form and content of a science. In company with a number of other and more peaceful examples, we may enlarge our understanding of what is contingent about the growth of knowledge, and what is not.

WEAPONS

From time immemorial weapons have been a product of human knowledge. The relationship became reciprocal. A great deal of the new knowledge being created at this moment is a product of weaponry. The transition occurred in World War II, and, in the West, was institutionalized by the new ways of funding research and development put in place in 1945–1947 in the United States.

Presumably this makes some difference to what we find out. Brains and equipment are dedicated to the production of knowledge and technologies useful in time of war. Our *Physical Abstracts, Chemical Abstracts, Biological Abstracts, Index Medicus,* and their on-line equivalents—our repositories of references to new knowledge—would look very different if we had different research priorities. That means that the *content* of our new knowledge is much influenced by the choice of where to deploy the best minds of our generation.

Outspoken people who urge us to find out more about living than dying deplore this distribution of research resources. But the picture that is suggested is rather like a menu: we cannot afford (or eat) all three of the entrees: meat, fish, and vegetarian. So we settle on one, but our choice does not affect the menu. Choosing meat today has no consequences for fish tomorrow, unless the restauranteur did not purchase enough fish, guessing we would go for meat again. But that defect can be cured in one more day, and the menu is restored. Thus one day we order up fibre optics communications that resist the electromagnetic pulse which wipes out standard signals systems upon a nuclear detonation. The next day, however, we could order up a solution to the death of the Great Lakes by poisoning, if we used comparable brains and comparable material resources.

I do not quarrel with the menu view, except that it deflects us from the menu itself. It implies that there are all those things out there in the world waiting to be known, and we choose which to know. But is there not the possibility that the very form of the menu may change, and in the case of knowledge, change irrevocably? May not new knowledge determine what are the candidates for future new knowledge, barring others that, in other possible human worlds, would have been candidates for knowledge? May not a direction of research determine not just the content of our *Abstracts,* but the very form of possible knowledge? There is a nagging worry that science itself is changed: not just that we find out different facts, but that the very candidates for facts may alter. In romantic but familiar terminology, we may live in different worlds for two different reasons. One is material. Our soaring triumphs and our poisons exuded by technology equally change the face of the material earth. So we live in a different world, thanks to our knowledge, from that of 1930, say. But we may also live in a different world because our conceptions of possibilities are themselves determined by new knowledge—a theme familiar from T. S. Kuhn.

One of my tasks in what follows is to provide examples of how the boundaries of knowledge are formed by the direction of actual knowledge. The boundaries of knowledge lie between the possible and the unthinkable, between sense and nonsense. We are creating these boundaries all the time. When so much knowledge is created by and for weaponry, it is not only our actual facts and the content of knowledge, that are affected. The possible facts, the nature of the (ideal) world in which we live become determined. Weapons are making *our* world, even if they are never exploded. Not because they spin off new materials, but because they create some possibilities and delimit others, perhaps forever. How are we to think about that?

THE CREATION OF KNOWLEDGE

I am thus concerned not with the use of knowledge but with its creation. My questions will often be abstract, compared to brute facts about our use of knowledge—such as the fact that the nations of the world have spent 17 trillion on weapons since World War II. My topic is, however, not unrelated to that. It arises from the fact that we generate much of our new knowledge in order to make better weaponry.

Where it makes sense to distinguish public from private financing, as in the United States, the bulk of public funds for research and development after World War II was dedicated to weaponry.[1] It is a rule of thumb, to which there have been notable exceptions, that the more talent and the more material resources devoted to an investigation, the more we find out, and the more quickly we find out. The sheer amount of investment in weapons research virtually insures that that is where much of our new knowledge is brought into being.

There is of course an important if obscure distinction between basic and applied research, so that R&D covers a multitude of practices. During the Cold War, the U.S. Department of Defense spent about 20 percent of basic research money on in-house laboratories, contracted out about 40 percent to private industry and the remaining 40 percent went to universities. This does not include nuclear weapons research, which comes primarily from the Department of Energy. It was a deliberate policy, regularly stated in congressional hearings, not to insist that all of this was "mission-oriented." In particular, the university workers should be allowed their heads, without too much direct control. All the same, the DoD would be paymaster and select the projects and the directions in which they should proceed. So when I speak of much basic research being devoted to weaponry, I acknowledge self-conscious policies of using military funding to create knowledge that is held to be both basic and not mission-oriented. Other nations have different policies.

Many would urge that the most successful growth of knowledge since 1945 has been in molecular biology, whose British, French, and American founders had, in the beginning, precious little of anyone's money. So much better endowed today, molecular biology continues on nonmilitary funding. There are indeed endless projects that rely on nonmilitary money, and many that survive on almost no money at all, Sometimes these are the best. But many of our momentous achievements, with innumerable peaceful spinoffs, were systematically created from military accounts. The laser is an example. We now think that almost anything can be made to "lase," yet quite likely almost no substance in the solar system ever "lased" until our own lifetime. This is a remarkable achievement. It is sometimes thought that lasers and Star Wars are a recent marriage in which a peaceful inquiry is put to military purposes. What could be more peaceful than using a laser for eye surgery, or totally transforming music reproduction in the form of compact discs? Yet the basic research leading up to the laser was not peaceful. It was made strictly on Department of Defense contracts, as a possible successor to radar and microwave technology. I shall give a few details below.

Despite the notable counterexamples, and despite the exciting peaceful spinoffs, much of our new knowledge has been made in the pursuit of new weapons. This conclusion raises vast issues both for morality and for policy. I deliberately avoid them. Many readers will be convinced that it is evil to spend the national treasure of brains and resources on new agents of death. The moral issues are philosophical and they concern science, but there is no reason to think that a philosopher of science—as the term is professionally defined in North America—will be well qualified to discuss them. Philosophy of science falls under metaphysics and epistemology—what there is and how we find out about it—while I have been citing familiar issues in ethics and policy. There are philosophers of science who write essays defending torture as an instrument of public policy and others who disagree, but qua philosophers of science, they are no better qualified to discuss such matters than is a truck driver or a Xerox repair person. Likewise philosophers of science should claim no more expertise on the ethical issues than, say the man with an office adjacent to mine, who is a classical archaeologist; less, perhaps, since he is the world expert on ancient Mediterranean archery.

Rather than discuss the ethical issues (about which I have strong opinions, but claim no rights as an expert), I shall bring to bear the most pressing and yet the most obscure of questions in contemporary philosophy of science. It takes us through the fashionable idea of the social construction of scientific facts, and the antique problems of skepticism and nominalism.

Before proceeding I should clarify what I mean by weapons. Our obsession with nuclear weapons tends to make us think of the bomb as the paradigm weapon. A moment's reflection reminds us that the essential part of weaponry is most often not what actually kills or destroys, but the euphemistically styled delivery system. The oft-invoked sturdy English yeomen who won the battle of Agincourt in 1415 did so not because of their arrowheads but because of their hi-tech longbows (and their deployment). The most brilliant military-scientific complex ever formed before the Manhattan project—Napoleon's group of mathematicians—was brought together to solve problems in ordnance, namely, how to ensure that the cannonball both traveled far and hit the target quite often. The Manhattan project is almost the only example in which the killing device—the atomic bomb—was the sole object of research, while the delivery system was a routine bomber on a routine flight. The Soviet triumph in atomic and nuclear warfare was not, as is commonly thought, the rediscovery of how to make the bomb, but the development of crude but unbelievably powerful rocketry. It is sometimes forgotten that the hardest problems to solve in Star Wars research involved not missiles and lasers, but guidance systems and fifth- (or later) generation computation.

When I speak of weapons, then, I include a whole gamut of military technology. Certainly I do not restrict myself to sensational weaponry in the news. I also have in mind the computational and artificial intelligence knowledge required for the windowless helicopter gunship, which would be very handy for counterinsurgency work. Even philosophers who write about the principles, the logic, or the statistics of perceptual systems can now get contracts for that sort of research.

FORM AND CONTENT

That old nag of a philosophical distinction, form and content, still has some life in it. I am concerned with the way in which the forms of areas of scientific knowledge are affected by their emergence out of military research. Much of this chapter will discuss examples that are not in general military, in order to bring some clarity to a question about the form of knowledge.

I might well have used in my chapter title the phrase "conceptual scheme" rather than "form of knowledge." Quine has, however, preempted the former term, meaning by it a structured set of sentences held for true. I think of a scheme of concepts as more like a framework for what can or may be true. By a *form* of a branch of scientific knowledge I mean a structured set of declarative sentences that stand for possibilities, that is, sentences that can be true or false, together with techniques for finding out which ones are true and which ones are false (cf. Hacking 1982, 1992b). Note that this is closely connected to Kant's idea of the origin of synthetic *a priori* knowledge. It is, however, very much of a *historical a priori,* to use the phrase of Michel Foucault. Thus what may be deemed possible at one time may not be held to be so at another. A form of knowledge represents what is held to be thinkable, to be possible, at some moment in time.

My account of a form of a branch of knowledge is deliberately nonjudgmental. One reader rightly noted that according to me, any set of declarative sentences, together with a Ouija board and a psychic, could count as a form of knowledge. Exactly so. It is a matter of historical fact that it does not. The various possibilities envisaged in the doctrines of the Trinity—including unitarianism—did constitute and do still for some people constitute a form of knowledge. They are not possibilities for me. Neither they nor their denials are part of my web of belief. In this respect my account parallels Quine's for conceptual schemes: he would have to take all the sentences declared true by a psychic, and accepted by her associates, as a conceptual scheme. Likewise for the Trinity or the transubstantiation of the host; likewise for the arcane sentences of Paracelsus.

As nuclear weapons have been the favorite topic in weapons philosophy, let me take for example the nucleus of the atom, without which there is no nuclear bomb. We can witness the coming into being of the nucleus, as a real possibility, in the years 1890–1912. I would say that in 1870 it was not thinkable that an atom should be constituted by an infinitesimally small concentration of mass in a void at whose outer limits are the remaining parts of the atom. It is true that Maxwell said there must be structure in molecules (by which he meant atoms), but Rutherford's atom was unthinkable. Certain possibilities did not exist for us, and only gradually entered the field as electrons came to be postulated and then known. Even when Rutherford did have the nucleus in 1911, he was very slow in talking about it, and did not at first much draw attention to it at the small congresses of the day. It really took him two or three years—not to countenance the nucleus as a fact, but to think of it as a possibility. The fact that the atom has a nucleus was less of a problem for Rutherford than to transform a form of knowledge in order to make an atom with a nucleus a possibility (and simultaneously a known fact).

Quine's aversion to modal concepts makes the idea of possibility, and hence of a form of knowledge, unattractive to him. Yet my procedure is almost alarmingly nominalist, verificationist, and positivist. I am speaking of nothing more than declarative sentences whose truth values can be determined and of the ways in which they are determined. The importance of the idea is that it gives us some general way to discuss the organization of constraints on directions of research, constraints that arise from a historical, *a priori,* absence of possibilities. Note that I do not say the exclusion of impossibilities. Slightly to abuse Wittgenstein's *Tractatus,* I would call something that is impossible *sinnlos,* while something that is excluded as unthinkable would be *unsinnig.* The Trinity, transubstantiation, and much of the work of my hero Paracelsus are, for me, *unsinnig;* so, I think, was an atomic nucleus, even for Maxwell, in 1870.

The notion of a frame of knowledge connects with many others that are at present well known, and it may even serve a useful deflationary purpose. Thus T. S. Kuhn, writing of scientific revolutions in a discipline or subdiscipline, speaks of changes in world view, even of a revolution leading us to live in a different world. A less romantic way to indicate the general idea is to say that the form of a branch of knowledge has changed; a new space of possibilities has emerged, together with new criteria for questions to ask and ways to answer them. Whether or not there *are* incommensurable forms of knowledge is a historical question, but at least the meaning of an assertion of incommensurability is moderately clear; there is no common measure between the possibilities that exist in one form of knowledge, and those that exist in another. Note incidentally that Donald Davidson's animadversions against the very idea of a conceptual scheme (a set of sentences held for true) do not so evidently apply to my notion of a form as a set of possibilities together with "methods of verification"—a crude but familiar label for a vast complex of ways for deciding questions.

Revolution sounds romantic. There are many more sedate ways in which the form of a body of knowledge can be historically determined, and might have been determined in other ways. I wish to steer away from grand talk of total conceptual schemes to more piecemeal things, and to steer away from talk of revolution to the manifold of complex ways in which not only the content but also the form of knowledge can be determined, altered, or constrained. I shall do this by a string of very different kinds of example, and which will include:

1. Early intelligence quotients
2. A now famous example from endocrinology
3. Detectors in high-energy physics
4. Lasers
5. Criteria of Accuracy for missiles

None of the examples is my own. I deliberately take historical case studies made by other people. The examples are not in general from weapons research, although I will from time to time point out military connections. Each represents a different way in which a form of knowledge can be molded. I wish to escape Kant's unifying idea that talk of the form of knowledge is talk of the one permanent form of knowledge. Since Hegel we have all become historicist, albeit in some cases kicking and screaming in resistance. Hegel denied the permanence of a form of knowledge, but not the unifying ideal. My talk of form is parasitic on a common idea of content. It does hold that at any time there are classes of possible questions bearing on some subject matter, and that ranges of possibilities change for all sort of reasons. A precondition for content is given by the form, the class of possibilities. But the determinants of these forms are multifarious. One of the reasons that the unity of science is an idle pipedream is that the forms of different bits of knowledge are brought into being by unrelated and unrelatable chains of events. Examples are needed to understand what this means.

Intelligence Quotient

The famous Stanford-Binet intelligence tests were set out along lines proposed by Alfred Binet, and then developed at Stanford University by Lewis Terman. Their authors were committed to the idea that biological characteristics should be displayed upon a Gaussian or Normal probability curve. I ignore the long and tortuous nineteenth-century origins of that idea. Binet devised questions which his subjects answered in such a way that scores shaped up on the familiar bell-shaped curve. The trick was to get a set of questions which, when answered, had this property. Terman, with his able female assistants who administered most of the tests, discovered that women did better on his IQ tests than men. Since women "couldn't" be more intelligent than men, this meant that the questions were wrong. Some of the questions that women answered better than men had to be deleted and replaced by ones on which men did better (Terman and Merrill 1937, 22–23, 34). This procedure fixed, for some time, the form of knowledge about intelligence. There were particular items of content, "how intelligent is Jones?" whose sense became fixed by the finalized method of verification, deliberately established by the investigator and his ideas. There also came into being certain synthetic *a priori* truths—and I mean this in exactly the sense of Kant. It became a synthetic *a priori* truth that women are no more intelligent than men. In passing, I emphasize that like so many of my other examples, this work had no military motivation or connection whatsoever. Yet war is always just around the corner. The Stanford-Binet test was legitimated and made both popular and semi-permanent by its use in screening American recruits in 1917.

In speaking of forms of knowledge, we appear to be close to questions of scientific realism. The flurry of discussions about scientific realism and anti-realism during the 1980s usually focused on questions of idealism, as to whether electrons exist, whether science aims at the truth or merely at instrumental adequacy, and so on. Those debates do not concern me here. Talk of a form of knowledge (despite owing much to Kant) takes one back to an earlier sort of realism, whose opposite is nominalism. The realist, in the sense that matters here, may well echo the first half of Wittgenstein's first sentence in the *Tractatus:* "The world is made up of facts." The nominalist retorts that we have a good deal to do with organizing what we call a fact. The world of nature does not just come with a totality of facts: rather it is we who organize the world into facts.

The nominalist controversy need not detain us. There is enough in common between nominalists and their opponents for the two sides to admit the phenomena I shall present. The attitude to the phenomena will be different, and the background talk about the phenomena will be different, but not enough for us to pause. For example, the nominalist says that the structure of the facts in my world is an imposition upon the world. The world does not come tidily sorted into facts. People constitute facts in a social process of interaction with the world and intervening in its affairs. Importantly, says the nominalist, forms of knowledge are created in a microsociological process. The person who believes the universe has a unique inherent structure will be offended by this description, but if attracted at all by the notion of forms of knowledge, may make use of an alternative background tale. It is this. The world is far too rich in facts for any one organization of ideas to trick it out uniquely in *the* facts. We select which facts interest us, and a form of scientific knowledge is a selector of questions to be answered by obtaining the facts. A rival, and if possible nonequivalent, form will elicit different facts. The facts are not constructed, although the forms of selection are. In what follows, it does not matter which variant of these two extremes you find most attractive.

It is easy to see that both nominalist and realist may give accounts of the IQ tests. The nominalist will say that IQ is an unusually clearcut example of a social construction. The realist may say that the Stanford-Binet test is objective (and confirm this along Stearman's lines by factor analysis) but equally agree this is just one way of ranking the intellectual abilities of people, attending to some aspects of (objective) intelligence.

Endocrinology

For a quite different example, consider the much cited book by Latour and Woolgar, *Laboratory Life* (1979, 1986). Acting as an ethnographer or participant observer in the Salk laboratories in San Diego, Latour was able to provide a first-hand account of a discovery in endocrinology that won a Nobel Prize. It seems a clear example of a discovery, one that even the most determined nominalist or constructionist must acknowledge. A certain hormone, or peptide, called thyrotropin releasing hormone (TRH), seemed to play an important triggering role in the hypothalamus, and thus be of importance to understanding mammalian endocrinology. Many laboratories competed but only two were successful, and they shared the prize.

Instead of completing a chemical analysis, both groups synthesized the substance TRH, which quickly became a standardly available substance manufactured by the Swiss drug company Roche. How could one talk about social construction, except in the trivial sense that social organizations did the laboratory work?

I take only some things from the Latour and Woolgar book, for some aspects of it seem to me to be far-fetched or dated. But here are some interesting things explained in greater detail in Hacking (1988a). There is almost no TRH in the world to analyze. Five hundred tons of pig brains had to be shipped from the Chicago stockyards on ice, in order to distill a microgram of TRH. And what was this TRH? It was a substance that passed certain assay tests. But there was no agreement on what the assays should be, and different labs had different assays. The winning labs "determined" the assays and so determined the practical criteria of identity for TRH. Second, when a certain peptide had been synthesized, and declared to be TRH, that was the end of the matter. The drug company that had sponsored much of the research patented and started selling synthetic TRH.

The question as to whether this really is TRH simply dropped out, with the skeptics turning their minds to other things. Synthetic TRH became a laboratory tool in its own right, and *Index Medicus, Chemical Abstracts* and *Biological Abstracts* now have it as a heading listing numerous monthly reports of experiments using TRH to investigate something else. (Do suicidal women become less suicidal when injected with the stuff?) Also much of the original interest, as having to do with mammalian brains, may have been mistaken, as TRH plays a role in the chemistry of alligators' stomachs. And so on: a whole research field is created, but, argue Latour and Woolgar, not because we simply revealed a new fact, which we use as a stepping stone to the next bit of discovery. Instead, a social sequence of events fixes TRH as "the" substance originally of interest, without it being clear that the experimental work had to conclude in this way. Indeed, once certain events occurred, there is no doubting the "reality" of the synthetic substance, TRH. Moreover, no one will ever challenge the system of assays that determine what TRH is, because it now defines TRH. Certainly the research work will not be "repeated"—who will collect another 500 tons of pig brains to distill a microgram of whatever it is?

This last, rhetorical, question may suggest that the example is all too easy. As a matter of sheer cost, no one is going to check out the original TRH experiments. The situation is anomalous, because, as Thomas Nickles (1988) has observed, after having established something, we constantly rework it, standing on the result but also modifying its initial appearance and relationships to other results. Scientific work is what he calls a "bootstrap affair." This fact is missed by accounts of research that emphasize the first appearance of a scientific fact, or the history that led up to it. Nickles objects that a good many philosophical histories or historical philosophies of science assume that a science gets to a result once and for all. Only the originating discovery, and its scientific or social antecedents, count. He calls these "*one-pass* models of scientific inquiry." In fact we constantly reconstruct and remodel results, apparatus, even phenomenology. Textbooks, which effectively delete the processes of discovery, are not to be thought of as distorting the history of science, but as being part of it. They participate in the ongoing reconstruction of a science.

So much is surely correct. In giving illustrations of how a set of questions and answers is put into place, I do not intend to foster one-pass models, but rather to emphasize how an entire field of inquiry may be formed. The example of TRH is closest, of all my five cases, to a one-pass view. The example may seem to concern the content of scientific knowledge rather than its form. The fact that TRH is a certain tripeptide would naturally be called part of the content of endocrinology. But facts do not just pile up blindly. They are used to determine the form of future inquiries. It is not just that the formula for TRH becomes a fixed benchmark in the science. The substance is manufactured and becomes an investigative tool, for it allows for certain new questions to be addressed, and certain new techniques to be deployed. This example has nothing to do with the idea I mentioned earlier, of form being a byproduct of scientific revolution. Nor is this case at all like the operations of Terman on IQ. Why I speak of form here is that certain issues have been closed off, and certain others opened up. An incredibly rare hypothesized substance is translated into an easily manufactured synthetic substance, which defines what is going on in your head. Nominalists Latour and Woolgar call this fact constructed. A realist need only say that among all the possible facts to be discovered in the endocrinology of the hypothalamus, this particular structure has been singled out and will determine the future possible structures to be discovered, shutting off others from the screen of possibilities.

Particle Detectors

A stronger example of the way in which the form of a science may be altered is well described by Peter Galison (1987). The bubble chamber was a chief detection device in high energy physics for over twenty years, although of course not the only one. It consists of liquid hydrogen under high pressure. When a very fast particle goes through this substance it releases bubbles, which serve as a track of the particle, and also of tracks of colliding particles, decaying particles, and so forth. It has the great merit that it is very "fast," allowing an enormous number of tracks to be observed in very short periods of time, whereas in older devices one had to wait a while between one good observation and the next.

The bubble chamber permanently changed high-energy physics. First, liquid hydrogen is incredibly explosive. That meant that a new level of staff had to be introduced into high-energy laboratories: safety engineers and the accompanying controls on scientists. Research physicists could not just wander around the lab any more. Second, for the first time very many more data were produced than any team of individuals could process. At first a new layer of observers, photographers, and counters was introduced, but this has all been replaced by magnetic tape and computer scanning. Moreover, in order that different laboratories could even understand their results, the tapes and their methods of interpretation had to be standardized. This was done at international conventions. A detecting device, the bubble chamber, did not merely enable one to detect what had not been seen before. It determined the form of the questions to be asked in high-energy physics in the world's laboratories. The inventor of the bubble chamber, Donald Glaser, was so appalled at the way his invention changed the day-to-day practice of physics that he left, Nobel Prize in hand, and took up molecular biology.

Most high-energy physics in recent years has had precious little military pay-off. In a larger view, it is widely agreed that the militarization of funding for science research in World War II was what made possible much big science such as high-energy physics. All the same, many who worked in the field felt that they were outside the arms race. Only in the early development of the atomic bomb were high-energy physics and weaponry intimately and necessarily related (for example, the first plutonium used to fire up Fermi's pile was made in the Radiation Lab and in the cyclotron at Berkeley, whose engineers also designed the calutrons at Oak Ridge for preparing enriched uranium). But even though I happily call the bubble chamber peaceful, it was made possible in its day by weaponry. It required a great deal of liquid-hydrogen technology, known as cryogenics, engineering done at very low temperatures and very high pressures. This cryogenic knowledge and material just happened to exist—in Colorado, where it had been prepared for Edward Teller's model of the hydrogen bomb. Teller's version was superseded, so it was possible to conscript liquid hydrogen technology and technicians for the first large bubble chambers.

This happy turning of swords into ploughshares has not been uncommon in high-energy physics. But the resources of equipment and of talent were made available out of Department of Defense funds, largely because of the old collusion between high energy and weaponry established during the Manhattan Project.

We can update this story of detectors to fit in with some themes of Chapter 3. One of Pickering's arguments for the contingency of high-energy physics relies on the fact that fundamental decisions about detectors were made in the 1970s. The bubble chamber was increasingly phased out, which meant that questions that might have been pursued faded away, and the new ones emerged. Pickering proposes to invoke Kuhn's concept of incommensurability at this juncture. The answers to new questions could not be compared to the answers to old ones, because the instrumentation that bears on the new questions is different in kind from that which bears on the old ones. Perhaps Pickering's strong contingency thesis is a matter of the contingency of the form of scientific knowledge rather than, as he presents it, a matter of its content.

Lasers

Lasers were perhaps the best known ingredient in the three-level Strategic Defence Initiative known as Star Wars. Despite this, one of the biggest investments was in the direction of advanced computation, in part as a way to subsidize the American computing industry into its next generations of computing power. While lasers may merely be the sensational tip of a vastly more complex program, it is still worth while telling a little bit about where they came from (Forman 1987).

Shortly before 1939, British scientists developed a primitive but valuable radar system for the detection of incoming bombers. (Even then they fantasized about "death rays" on the side.) Radar was, for a short time, as close to a defensive weapon as you could imagine. It quickly became used for offensive purposes, for example in locating enemy warships and in particular submarines that needed occasionally but regularly to surface. Throughout the Second World War there was intensive development of numerous microwave techniques, which continued unabated after the war. A number of projects were started which were aimed at producing exceedingly stable and reliable high frequency emissions, and the solution gradually proposed was to use artificially stimulated resonances of molecules themselves. Work on the maser was entirely funded by the Department of Defense. So were the two earliest programs to construct a laser (Harold Townes at Columbia and Gordon Gould at TRG Inc.). In the three years following the first demonstration of lasing, the DoD dumped 100 million 1985 dollars into research. Private industry also quickly responded, and soon was putting more into laser R&D than the DoD. It was, however, the DoD that gave us this phenomenon—a remarkable gift; for although the phenomenon of lasing (unlike masers) is becoming ubiquitous all over the industrialized parts of our planet, it existed (with all probability) nowhere in the solar system before 1950.

Peaceful applications of lasing are legion. Moreover, it will long continue to be a topic of profound basic research, particularly as it is an unusually accessible and manipulable instance of a nonlinear process. Assuredly we should express gratitude for this gift of the Department of Defense.

Why should I group it under the heading of a form of scientific knowledge? Let us suppose that there has been a pretty steady weapons thrust underlying laser research, a thrust that brought us to Star Wars research.

Does this not represent simply the steady investment of public funds in military research, churning up new discoveries that may have military application, and certainly have peaceful ones? On the form/content spectrum, is this not squarely on the content side?

I shall give two connected answers, one practical, and the other embedding that answer in a current philosophical tradition. The practical answer (given here from a realist, or inherent structurist, standpoint, but susceptible to nominalist rewriting) is that in the development of postwar physics there was no prior inner necessity for lasing to be discovered. There are endless aspects of molecular structure on which to work. The choice of problem was directed by the military. That done, we had another benchmark situation. This fundamental discovery served as a "paradigm" of inquiry—not due to any Kuhnian scientific revolution, but because other fields of questioning were screened off by this monumental success. For a substantial period of time to come, a wide range of possible questions will be formulated according to this paradigm.

The significance of this commonplace can be partially understood by connecting it with Lakatos's (1970) notion of a research program. An ordinary research program is a familiar beast, often described in an initial proposal asking for money from a patron. We propose to do this, this, and this, and if we are lucky we'll do that, and then try for such and such. A research program is pretty specific, should be flexible, and is finite. It may be replaced by another successor program in three years or three months. Lakatos's research programs are quite different. His own examples include ones that last a century and were driven underground, forgotten for decades while they lay fallow in a field of counterexamples. His research programs have a structure of positive and negative heuristics, of hard cores and protective belts, they may be progressive and degenerating, both theoretically and empirically.

An ordinary research program, as I understand it, is an inquiry that takes place *under* a form of knowledge, although its upshot may change that very form. There are certain questions, and certain ways of trying to answer them. When the program first surfaces as a proposal for scrutiny, it is supposed that the questions are intelligible and their answers at least partially attainable by the proposed techniques. The aim of a program is to increase the content of our knowledge and its uses. A Lakatosian research program, on the other hand, is not so far from my idea of a form of knowledge. Part of Lakatos's idea of positive and negative heuristics is that of the questions that can be asked, and those that cannot. Lakatos would have resisted my word "form" with its Platonic and Kantian overtones. But Plato and Kant were for fixed forms within a unified scheme; it will be clear from my previous examples that I markedly am not. I propose even more ways of changing programs than Lakatos ever got around to discussing. I am not unhappy to think of a Lakatosian research program as one of the ways in which to come to grips with my groped-for concept of form. I would not identify the two, any more than I would identify Kuhnian revolutions with the creations of forms of knowledge. This is because I want a flexible and many-valent concept. I do not think there are many exactly Kuhnian revolutions or exactly Lakatosian research programs in the history of science. The sciences have got on very well, much of the time in many places, without either notion being instantiated.

At any rate, it will now appear that in my thinking, Lakatosian programs and ordinary programs are fish and fowl, oranges and apples, or perhaps as different as fish from apples. But radar-microwave-maser-laser-SDI gives me pause. This is certainly no short-term program, written up in a few proposals and funded by the U.S. Army Signal Corps or whomever. It needs little stretching of Lakatos's own definitions to see this development as the working out of an identifiable research program, starting, indeed, with the only partly jocular thought of the British pioneers thay they might devise a "death ray."

Yet such stretching of Lakatos would belie some of his own intentions. He wanted research programs to be part of his philosophical theory of a purely internal, autonomous account of the growth of knowledge. Political, social and psychological factors were to be excluded. A dominant feature of "the laser program" (if there was one) would be that, despite its endless civilian spinoffs which burgeon apace today, there was one and only one major paymaster, the Department of Defense. That is, there is an entirely external account of what directed the program and got it moving.

Missile Accuracy

Here I shall allude only briefly to Donald MacKenzie's (1990) book on missile accuracy, in part because in certain ways it so resembles our first example of IQ.[2] At first blush it may seem that missile accuracy is an entirely objective concept; the missile either hits the target, or it does not. On second thought, it obviously isn't "objective" and that for several reasons. First, as in archery, accuracy must be graded, with top marks for the bull, and diminishing marks for increasing deviation from target. The grading depends upon the point of the exercise. If killing is the aim, the warhead will determine part of the measure of accuracy. For example, in a defense of Western Europe against an imagined tank attack, consider two missiles. *A* carries a relatively small amount of conventional explosive; *B* is a very low-yield fission bomb. To be "accurate," *A* must detonate very close to the tank that is its target. The constraints of accuracy on *B* are different; it can be somewhat off the center of a tank battalion to wreak havoc. On the other hand, in the jargon of the army, small towns in Germany are only a kiloton apart, and if part of the object is to be of some help to the locals, the missile *B* should be as close as possible to half a kiloton away from any village. Problems of accuracy for *B* are a good deal harder than for *A*; luckily they become increasingly moot with improved "missile accuracy" for *A*-type missiles. Evidently questions of missile accuracy become more complex for strategic as opposed to tactical nuclear weapons.

So there is a problem in defining how close one is to a target. The second problem is that one is not talking about the accuracy of an individual missile (which is fired only once). One is concerned with a type of missile, and missile accuracy becomes a statistical concept which is open to a good many interpretations.

MacKenzie argues that there was once an extensive debate on missile accuracy which has gradually stabilized into a set of measures and comparative standards. Every manufacturer and every branch of the armed services had its own standard of missile accuracy, often giving wildly different characterizations of the relative "merits" of different missiles. By what the nominalists call a microsociological process, consensus has been reached. This consensus determines in part the very construction and design of missiles (because you have to achieve accuracy within the designated limits, whereas another measure of accuracy would have called for a different design). It matters to arms control negotiations and much else. There appear to be many formal comparisons between this example and that of IQ. In crucial cases, answers to the question, "is this missile more accurate than that one?" are determined by the assay criteria on which the community has decided. These become part of the form of possible knowledge, defining the "content" answers to questions that at first seem independent of any "form."

These five examples serve to put some flesh on the skeletal idea of a form of scientific knowledge. It is all very well to say a form is a structured class of sentences that are all capable of being true or false. That is but to pose a question; namely, how do such classes come into being and how are they changed? My answer is manifold. It includes deliberation, as in IQ or missile accuracy. It includes the establishment of assay techniques as definitive. It includes the making of a substance synthetically that defines a part of nature. It includes the creation and standardization of detectors. It includes research programs and the external forces that give them direction. It does not, of course, exclude Kuhnian revolutions. Nor does it exclude that most general kind of form of knowledge that Michel Foucault called an *episteme,* The roots of the present chapter may be detected in some of Foucault's ideas of what makes positive knowledge possible.

FORMS OF KNOWLEDGE

Alas, I have no simplistic conclusions. My original aim in discussing weapons was to connect traditional philosophy of science and the weapons research question. I attempted to open a debate, not to close one. The work is in part a response to three commonplaces. (1) We have enjoyed remarkable spinoffs, of great benefit to humanity, from weapons research and military funding. (2) The human race learns more in times of war and of rumors of war than at other times. (3) Knowledge can be put to good uses or evil ones; the use of knowledge is a matter of public policy, not science.

I have been at pains, in my examples, not to deny the first assumption. It is not that I asserted it, but I provided illustrations, some unfamiliar, that could be used to defend it. However, the claim about spinoff knowledge is not particularly germane to my concerns. Insofar as there is a viable form/content distinction, (1) is about the spinoffs from particular contents of knowledge.

The second statement, about the fertility of research in times of war, is connected by its proponents to the spinoff doctrine, (1). However, it seems to be false. It is true that prosperous wartime and war-preparation economies provide ample funds and motivation for discovery. It is true that wartime shortages also invite invention, such as that of artificial rubber (after loss of colonial Malaya) or sugar from sugar beets (after the effective loss, in the Napoleonic wars, of the equally colonial French West Indies). But it requires some talent to list war-related discoveries in the warrior nations of Europe between 1914 and 1918. Rocketry, nuclear power, and microwave technology are among the adventures accelerated during 1939–1945, but the greatest scientific achievements of our era, in terms of knowledge, are surely (to take the interwar years) the new quantum mechanics of mid-twenties Weimar Germany, or the early postwar triumphs of molecular biology (made, in many cases, by men who had wasted six years in "war work"). Platitude (2) isn't true, but if it were, it too would be about content, not form.

I have no doubt that in many respects the third proposition is true: in the case of nuclear weapons, the few powers great and small who own them could without inconvenience eliminate them in a few years. It is a political choice, which may be wise or foolish, not to do so. Should the scientists, the creators or possessors of the *knowledge* that makes the weapons and the delivery systems possible, be part of the political scene? That is an important issue that I have forsworn here.

Like (1) and (2), the third commonplace operates at the level of individual matters of fact. There is this knowledge, crafted by human minds and hands. This knowledge may then be used by other minds and hands for good or evil. This statement (3) and the corresponding ethical problems are stated, quite appropriately, at the level of matters of fact, of content.

All three commonplaces are governed by that very picture encapsulated by Quine, of knowledge as a conceptual scheme, of a set of sentences held for true. That cheerful empiricist picture of a holistic structure of sentences says nothing of questions, except in the form of whether items in the scheme are, after all, true. It says nothing of which questions we are able to ask at a time, and of how the arrangement of possible questions can be changed. It says nothing of how the scheme will be altered by a radically new invention (the bubble chamber) or how making a new substance sets up strategies for attack on old questions in a new way. It says nothing about how a program can turn radar into the laser and give us the Strategic Defense Initiative (as well as many goodies on the side).

The conceptual scheme-picture is one of autonomous knowledge living its own life, with its bosom buddies, the scientific investigators. The form of knowledge-picture is one that admits that possibilities are constrained in a manifold of complex ways at a particular time. What we can think of, what we want to ask, what we want to do as investigators is a historical event. It is not rigid, but neither is it altogether fluid.

Copper is malleable and ductile, but you can't do *anything* with copper; likewise, forms are malleable, but still operative. We have long had the fantasy that attending more closely to the forms of knowledge will somehow be liberating. That fantasy is not automatically to be dismissed when it is introduced into new and parlous territory. It is to be transformed into something more than fantasy, and one way to do that, in my opinion, is to get a fairly rich diet of examples.

I would altogether deplore the inference that forms of knowledge connected with research primarily funded by the military are wittingly created by those who are responsible for weapons research. Such ideological paranoia is absurd, if only on the ground that, contrary to what I write, the concept of a form of knowledge may be either inexplicable or when explained, empty. I am more concerned that we have no idea of what we are doing in the overall directions of our conceptions of the world. There is no monolithic military conspiracy in any part of the globe to determine the kinds of possibilities in terms of which we shall describe and interact with the cosmos. But our ways of worldmaking, to repeat the phrase of Nelson Goodman, have very often been funded by one overall motivation. If content is what we can see, and form is what we cannot, but which determines the possibilities of what we can see, then we have a new cause to worry about weapons research. It is not just the weapons—we can dismantle them in a few years with good will—that are funded, but the world of mind and technique in which those weapons are devised. The forms of that world can come back to haunt us even when the weapons themselves are gone. For we have created forms of knowledge which have a homing device. More weapons, for example.

Chapter Seven

ROCKS

Science studies, sociology of scientific knowledge, science and technology studies: these are where the action has been in the philosophy of science over the past few years. I do not mean to belittle specialist studies of quantum mechanics, space and time, systematic biology, neurophilosophy, or questions of cause and effect, theory and experiment, probability and induction, and much else. But if we ask what has played the thoroughly lively role, in the republic of letters, of Thomas Kuhn, or Karl Popper, or John Stuart Mill, or Francis Bacon, in recent years, the answer must be science studies.

This is not a majority opinion. Most philosophers of the sciences resent the trendy and iconoclastic character of social studies of knowledge. Some rise to anger that disguises terror. I myself find the fuzzy dragon of science studies rather endearing, but I also retain a lot of respect for more traditional philosophical thinking about the sciences. Here I want to show how well the old-time philosophy fits a current example of science in action, but also to illustrate some insights from social studies of knowledge.

One way to do philosophy is to take a careful look at some corner of the world. That ensures some rigor, but accuracy must not be myopic. The example must illustrate, and serve as a parable for, a general point of great interest. My choice here is up to the minute and into the future. No one knows how the story will end. The facts have not stabilized. But it certainly is an exciting story. We could be talking about the origin of life, maybe even life before Earth. And, like some of the most fascinating science, it starts with the dullest, most ordinary stuff imaginable. It starts with dolomite, a substance similar to limestone (which is mostly calcium carbonate), except that it is mostly magnesium carbonate.

First let us have a sketch of the science (ending in science fantasy), and then a philosophical commentary. The science itself breaks into two parts, old science, and new science. The old science is sedimentology, where I rely on McKenzie (1991), who herself made good use of von Morlot (1847). This is an old story. But also a new one. The new science involves nanobacteriology, the study of bacteria a thousand times smaller than the bacteria normally studied by bacteriologists—many of whom suspect there are no nanobacteria. (*Nano-* denotes 10^{-9}, or one billionth, just as *micro-* denotes 10^{-6}, or one millionth, and *milli-* denotes 10^{-3}, or one thousandth.) Commonly studied bacteria are around a micrometer in diameter, while nanobacteria are about a nanometer in diameter.[1]

SEDIMENTOLOGY

Dolomite Recognized

In 1791 a French geologist, Déodat de Dolomieu (1750–1801), identified a distinct type of limestone in the Tyrolean Alps. He was a bit of an adventurer, one of Napoleon's team of scientists in Egypt, who became a leading figure at the Ecole des Mines in Paris. Next year Nicolas-Theodor von Saussure, scion of the most eminent family of Swiss geologists in those revolutionary times, named the stratum and the mountain region after Dolomieu. That ensured immortality for the Dolomieu family name.

Already we have an object lesson in the history of science. Dolomieu's deposit of dolomite had been identified as magnesia limestone twelve years earlier by a Tuscan mineralogist and metallurgist, Giovanni Arduino (1713–1795). His was the more remarkable feat, because the element magnesium itself had only just been recognized. Moreover Arduino proposed what has remained the fundamental hypothesis about dolomite formation. The magnesium carbonate is not aboriginal; it is a replacement compound that results from the substitution of magnesium for calcium in ordinary limestone.

As it happens, Saussure analyzed dolomite erroneously in 1792, and concluded that it was high in aluminum and had almost no magnesium at all. Dolomieu agreed. It took well over a decade to get the analysis right (Zenger at al. 1994). Dolomieu's superficial description of the mineral was sound, but only Arduino understood what he had his hands on, namely a magnesium compound.

Here we have one of those doublets in the history of science that the pioneering sociologist of science, Robert Merton, studied so intensively: independent discovery of the same phenomenon at about the same time. Thomas Kuhn (1977), always dedicated to the role of theoretical understanding in science, wrote an important paper on multiple discovery, using the case of oxygen. Scheele, then Priestley, then Lavoisier, in that order, all captured oxygen in a flask. But the palm of honor belongs to Lavoisier, in Kuhn's reckoning, because only he knew what he had in that flask. The relation between Arduino and Dolomieu is almost the opposite. Arduino had the better theoretical grasp of the substance, but Dolomieu got the glory. In fact he recently got a book of essays in his honor (Purser at al. 1994), including a piece titled, "Dolomieu and the first [*sic*] description of dolomite" (Zenger at al. 1994). On Kuhn's theory about who discovered what, it was Arduino who discovered dolomite, not just because he got there first, but because he knew what dolomite *is*, namely, a magnesium carbonate sedimentary rock that is the result of magnesium replacing calcium in limestone.

Why did Arduino think that dolomite results from replacing calcium by magnesium? I do not know why he realized that the key element is magnesium: that is a question for historians of geology. We can readily see why he thought dolomite is the upshot of a replacement process. Dolomite is porous. We would not expect that, if the sedimentary layer had been deposited as dolomite in the first place. So why is it porous? Because some substance or substances have replaced calcium during or after deposition, and the replacement compound is less voluminous than calcium carbonate. Having grasped that the substance is magnesium, one next realizes that magnesium ions are plentiful in sea water. Conjecture: dolomite sediments must be the result of a chemical reaction at the bottom of ocean beds. That seems to have been Arduino's original insight.

Arduino was not well networked into European geology. His understanding lay fallow. Geology followed a red herring. The replacement idea did catch hold, but went off in the wrong direction. In 1824 Leopold von Buch (1774–1853), in his day perhaps the most distinguished geologist from the German-speaking lands, made an exacting study of the Tyrol, and suggested that the eruption of augite porphyry could have provided the source of magnesium for the carbonate. He had a vision of hot magnesium gases interacting with the limestone. He even held that the intrusion of porphyry, accompanied by dolomitization of limestone strata, was responsible for the genesis of mountain chains. At his most enthusiastic he urged that it was the cause of *all* mountain uplift (R. Laudan 1987, 194–96). Thus dolomite began its career as part of a story about the origin of almost everything rugged. Not the placid tale of deposition at the bottom of the oceans, but of catastrophic intrusions from below. The failure of that vision is part of the classic history of geology.

Dolomite Formation by Experimental Chemistry

The first experimentally testable idea about dolomite was put forward in 1845 by another Swiss, Wilhelm Haidinger. He suggested that limestone is attacked by a solution of epsom salts (of which the key molecule is magnesium sulphate). The magnesium is carried away in a solution of gypsum (calcium sulphate). It was fairly easy for Haidinger to produce the reverse reaction, in which dolomite and gypsum turned into limestone and epsom salts. Since we are talking about harmless household chemicals, you may well be able to do this in your bathtub, turning dolomite into limestone in the presence of gypsum, producing a warm solution rich in epsom salts and a precipitate of limestone.

Haidinger could de-dolomitize in his laboratory, and he proposed that this process occurs naturally on the surface of the earth. But the reverse chemical reaction requires heat and pressure. Dolomitization could, he thought, occur only in the earth's interior. A Swiss geological chemist, Adolphe von Morlot, immediately took up the challenge. Apparently he produced dolomite at 250°C and 15 atmospheres. Unfortunately this tour de force of experimental geology did not help explain the great masses of dolomite found near the surface of the earth. They appear to have been formed at ordinary earth-surface conditions, and not under high pressure or temperature.

The Dolomite Problem

The central dogma of historical geology is Lyell's proposition that the earth has been fundamentally the same since it was formed. The long name for this doctrine is substantive uniformitarianism (Gould 1977). Dolomite seems to provide a counterexample. The early earth abounds in dolomite giants. But today it is not being formed in significant amounts. That is the enigma of dolomite, simple and sharp. What conditions prevailed early in the history of our planet, that allowed for the formation of masses of dolomite? What seemed like a fussy little question for sedimentologists is now assuming grander proportions. We have to find conditions that obtained early in the history of our planet, which do not obtain today. Does dolomite provide a window through which to look back on our past?

The presuppositions of the modern dolomite problem are, like everything else in this story, easy to state. First, the massive dolomites have been produced by magnesium replacement in calcium carbonates. Second, sea water is the ideal source of dolomitizing solutions because it has a high concentration of magnesium ions. Third, there were immense dolomite formations long ago, but they are minuscule in the geologically recent past.

Our story provides a nice example of how a fundamental belief, in this case uniformity, can lead to the creation of what, in retrospect, look like mythical data. The next best thing to uniformity is gradualism. From 1909 until 1987 it was pretty much agreed, on the basis of four independent sets of data, that the rate of dolomite deposition had not only gradually decreased throughout the relevant geological epoch, but that the rate of decrease was linear with time. So one ought to look for a change in, for example, the composition of sea water that was regular and linear with time (as if the oceans aged at a linear rate). Perhaps there was less and less magnesium in the ocean, a gradient that would explain the decreasing production of dolomite. In 1987 two geologists re-analyzed and re-presented the existing data in such a way as to show that this conclusion is spurious. "If this re-evaluation is correct and the formation of massive dolomites is not a time-dependent process, another dolomite myth must be discarded" (McKenzie 1991, 44). Not because of new observations or experiments, but because of new eyes re-examining old data.

When pleasantly simple myths are discarded, the result is a mess. Endless problems need to be addressed. There is some sort of magnesium cycle in the oceans, which is not well understood. Sometimes magnesium ions are plentiful, sometimes not. That is a problem of ocean chemistry. There is a related problem of hydrology. When magnesium is around, what pumps it to suitable sites under suitable (and totally unknown) conditions?

Dolomite Is Being Formed Now

The claim that dolomite is not being formed on the surface of the earth in the present geological epoch turns out to be another myth, although one that is closer to the truth. The truth is that it is not being formed anywhere that people find it nice to be.[2] It is being formed in regions hostile to almost all forms of life. The sabkha regions of arid sands and shallow seas on the coasts of the Persian Gulf. Salt lakes, mud flats, and deep-sea anoxic environments investigated by the deep-sea drilling research vessel. Or foul coastal swamps in Brazil. That does make one think. Maybe what is peculiar to dolomite is not only mineralogical but also biological: it just grows in regions in which nearly all modern life forms are severely stressed.

Models

What about the formation of dolomite in sea water? Hydrologists make models of what might have been happening, and then see if on the one hand they are theoretically feasible, and on the other hand, whether real-life events occur as in the model. There is, for example, a model of mixing between fresh and sea water. It leads one to expect regions that are supersaturated with dolomite, but undersaturated with calcite. So dolomitization of the calcite sediments should occur. Well, there are lots of mixing zones around the world, and this does not appear to happen.

One place that modern, although minuscule dolomitization, does occur is in the sabkha environment on the Trucial Coast of the Persian Gulf, at Abu Dhabi, where incredibly arid land abuts shallow seas. An evaporation "pump" could work here. Ground water rises up through calcium-rich sediments by evaporation at the surface. Marine water flows in to replace ground water, especially during periods of extreme tidal flooding. Yes, small amounts of dolomite are being formed right now. By radioactive strontium tracing it has been shown that marine magnesium ions do move to the supersaturated layers where dolomitization takes place. "In this sabkha model, evaporatively modified sea water is the dolomitizing fluid and evaporative pumping is the hydrologic mechanism, the magnesium pump, which draws the fluid upwards through the saturated sediments" (McKenzie 1991, 47). Can this model account for the giant dolomite formations of the past? There is a good deal of room for controversy. One has to imagine such a specific set of circumstances that the sabkha model seems, to many people working the field, to be implausible on the grand scale required.

We are in the realm of speculation. We are talking about the life of the planet. It is uniform in the grand scale of things, perhaps, but not constant. Ice ages are known to every schoolchild. Glacial intervals are frequently associated with long arid periods. Maybe we could have pumping of sea water through carbonate platforms, driven by climatic changes. There is some evidence that this is happening in the great Bahamas undersea platform.

McKenzie concluded her survey paper of 1991 with a nice aphorism: "The Dolomite Problem is a centuries-old problem which changes its face to fit each new generation of biologists" (p. 52). She also described the issues as forming a "labyrinth." Now let us pass to a new but not-so-new inhabitant of the labyrinth.

NANOBACTERIA

Bacteria have long been a player in sedimentology. There are many roles for microbial agents. Suppose for example that a certain ion inhibits a certain chemical reaction. Bacteria that get rid of the ion could thereby encourage the reaction. Some seventy years ago G. D. Nadson performed experiments with anaerobic sulphate-reducing bacteria taken from anoxic sediments in a salt lake in Russia. He was able to precipitate carbonates, some of which were magnesium carbonates. Perhaps sulphates inhibited dolomite formation, and the bacteria, by eating the sulphates, paved the way for more dolomitization. He suggested that we need to understand the role of bacteria in this phenomenon in order to solve the dolomite problem.

Much more recent work made the reduction of sulphate ions seem increasingly relevant. Investigations of the continental margins of Baja California and the Gulf of California showed dolomite in active formation. Lack of sulphate ions in these strata made sulphate ion-inhibition a contender for our inability to reproduce dolomite formation in earth-surface conditions in the laboratory.

The Road from Rio

It is a myth, we said, that there is no dolomite formation near the earth's surface today. Crisogno Vasconcelos, a Brazilian graduate student at the Eidgenössische Technische Hochschule (ETH) in Zurich did his thesis research on a coastal lagoon near Rio de Janeiro. It is called Lagoa Vermelha. That name, and the word "lagoon," sound romantic to those of us brought up on *Treasure Island* and the like. In fact this is a hot, smelly, saline swamp. It is nowhere deeper than two meters; it is less than two square kilometers in area. Although we think of this as a humid region, Lagoa Vermelha is technically semi-arid, with a major net evaporation loss in the dry season. The adjacent sea is hot and shallow, and already has a high concentration of salts due to evaporation. Since the land around the lagoon dries out, salty sea water seeps through the land into the lagoon (which is also sometimes flooded by the sea itself, during storms). So here we have a disagreeable environment (for people) which renews itself every season.

Vasconcelos and his assistants extracted cores of black sludge from the lagoon. I do want to emphasize that the procedures used throughout were wonderfully simple. Black plastic tubes, of the sort used for cheap plumbing almost everywhere in the world nowadays, were pushed into the mud as far as they would go, namely about two feet. The stuff at the end was freeze-dried for further study. Yes, there was dolomite there, and yes, there were sulphate reducing bacteria.

The next step was to culture the sludge. In the first instance this was done at the Idaho National Engineering Laboratories, a world center for bacterial culture. After the mud tested positive for sulphate-reducing agency, it was added to growth medium and clean quartz crystals. It was kept for a year at −4°C in a refrigerator, and finally taken back to Zurich for study. When opened, the vials stank of rotten eggs (hydrogen sulphide gas). Something had been at work taking out the sulphate ions.

Results: (1) a scanning electron microscope's secondary image showed the quartz grains covered with a knobbly coating "that appears to be colonized by subspherical nanobacteria" (Vasconcelos et al. 1995, 221). (2) An energy dispersive X-ray spectrum showed high intensity peaks for calcium and magnesium, and X-ray diffraction was "consistent" with being an ordered (crystalline) dolomite. (3) The crystallographic character was further studied by transmission electron microscopy. "Together, our analyses provide conclusive evidence that the bacterial production of dolomite can be achieved in low temperature anoxic conditions in a relatively short time."

Nanobacteria are something new (and old). And small. Ordinary bacteria are about a micrometer, or a millionth of a meter in diameter. Nanobacteria, if there are any, are only one thousandth that size. Nanowork is now becoming possible with recent technology, hence the expression nano-engineering, coincidentally pioneered in Zurich by IBM's research wing there, with the invention of the atomic force microscope. Nanobacteria are highly unorthodox. It is by no means agreed that there are any.

For some years Robert Folk (1993; Sillitoe, Folk, Saric 1996) has been urging that there are such things, closely associated with carbonate sediments. That is a highly controversial claim. For the dolomite, the argument is largely based on the appearance of subspherical minute bumps. "The nanobacteria appear to be encrusted by nanocrystals of dolomite, which apparently precipitates on the outer surface of the bacterial bodies. Some of the nanobacteria may be in the process of reproduction cell by division." Or maybe, a skeptic will add, those are not bacteria you are observing at all—certainly not nanobacteria, because there are no such little creatures in existence. Such absolute skepticism may be a hard thesis to defend in the future, because nanobacteria are, one might say, in the air, or rather, in the blood. Two Finnish scientists, Kajander and Çiftçiglu (1998), have found what they call nanobacteria in the blood of humans and cows, and have shown they are genetically similar to common normal-size bacteria that infect those animals. These minibacteria (not, in fact, quite as small as a nanometer, but nearly) erect little mineral coatings to protect themselves; the authors speculate that these coatings may be the seeds that grow into kidney stones.

The next step for the students of dolomite must be to do what the Finnish team did for nanobacteria in the blood—investigate the genetic structure of these alleged bacteria. We wait. But let us see where speculation takes us. McKenzie and Vasconcelos "further suggest that the bacterial production of dolomite may have been very important in the past, especially in the very early Proerozoic epoch, when three times as much dolomite as limestone was produced."

We are now talking about a period not far from the formation of the earth's crust. That means that we are now—and this is not yet stated in *Nature* or the *Journal of Sedimentology*—talking about very early life forms indeed. Let us go further. The potential for genetic change in organisms this small is thought to be minimal. The bacteria in Lagoa Vermelha may be, essentially, among the first living creatures that appeared on earth.

Why stop there? You may recall the chunk of rock from Mars that created a public relations splash in 1995. Some scientists thought they found signs of primitive life fossilized on the rock. The signs are remarkably similar, in some respects, to the alleged nanobacteria. In fact the speculations about Martian life have now fallen into disfavor, but they are not entirely abandoned.[3] Just suppose that the origin of the nanobacteria of the type found in Lagoa Vermelha is extraterrestrial . . .

OLD-TIME PHILOSOPHY OF THE NATURAL SCIENCES

Let us now turn to philosophy, and argue for four theses. (1) Traditional philosophies of the natural sciences give useful descriptions of the research just described. (2) Many of the facets of the natural sciences that have been emphasized in recent social studies of science are well illustrated by the story of dolomite. (3) The three sticking points of Chapter 3—contingency, nominalism, and stability—can provoke the usual arguments. (4) Rocks are just as problematic as Maxwell's Equations for the fundamental philosophical issues that underlie the social-construction controversies.

One might like to cut a dashing figure, and use these theses to propound paradoxes. On the contrary. The upshot of this little case study is bland. Talk about the social construction of dolomite, or of the origin of life, if you will, but the dragon of construction has been defanged. The curmudgeonly troll of hyperbolic scientific realism may go on grumbling in his dank cellar, but he will no longer shake the whole mountain in his rage. "Oh, but you have just chosen an easy example." No, it is a very hard example, harder than electrons and equations, precisely because it is so commonplace. It is a typical example of what many scientists all over the world are doing, right now.

Small Science

Before any philosophy, let us look at the scale of the research just described. In ambition it is potentially enormous. But it is small science. A dominant theme in recent historiography of science has been big science. It was Vincent de Solla Price who announced it. One fine example of big science studies is Galison's (1997) book, whose size (over 1000 pages) is worthy of its topic. Science and technology changed in innumerable ways during the Second World War. The Manhattan Project became an institutional model for postwar big science. The National Aeronautics and Space Administration followed suit. The Human Genome Project is in the same mold. Big science is a major player, and history, philosophy, or sociology of science cannot ignore it.

Some scientists are skeptical about the potential of big science for genuine innovation, none more so, to my knowledge, than Freeman Dyson. He was active in operations research during World War II (the beginnings for what Pickering 1995 calls "cyborg history," and hence for the "regime" of big science). He was one of the handful of physicists who brought quantum electrodynamics into being. He has long urged that the major novelties in human discovery will not spring out of the great laboratories—prestigious, well-funded, with their pools of brilliant talent. The really new ideas will come from the scientific fringes, undernoticed, forced by the exigencies of weak financing to improvise and to think, rather than to deploy vast armies and treasure chests of materiel. Small science, he thinks, will be the source of the rare stunning novelty that changes our vision of the world. To exaggerate the thesis: big science is bound to be what Kuhn called normal science, while revolutionary science will, from now on in, occur on the fringes.

I have just told a story of small science on the periphery. The speculations of Judith McKenzie may simply not pan out. Her more extravagant conjectures may be thrown out as rubbish. But there is no doubt we are on the margins. The project was the work of a Brazilian graduate student with a woman professor. She is, indeed, at a fine, if rather conservative, institution—the ETH—with more Nobel prizes to its credit than the whole of Canada. But the Swiss institute is (frankly) only just beginning to consider women engineers and scientists as participants. McKenzie's curriculum vitae is not that of the hotshot, even in the fringe science of sedimentology. We have a Brazilian graduate student extracting sludge with some plastic plumbing you can buy at your local hardware; at present his bacteria are being incubated in an ETH refrigerator where laboratory workers keep their lunches. If the work of this group were to lead to a striking role for nanobacteria in the early history of the earth, it would be a coup for small science, of exactly the sort that Dyson predicts.

Philosophies of the Sciences

Some philosophers would say that the old times, perhaps the good times, are the days when Thomas Kuhn, Imre Lakatos, and Paul Feyerabend were the talk of the town. For others, the old times, perhaps the bad old days, were the heyday of Karl Popper, or the logical positivists. Perhaps the really good old times are when the philosophy of the sciences was inaugurated—or at any rate, when the name "philosophy of the sciences" first came into use. That was in 1840, by William Whewell (1840, I, 2), who also is said to have invented the very word "scientist."

Needless to say, philosophical accounts of the sciences did not begin in 1840. In the English-speaking world, there have been two great traditions. One is the inductive method of Francis Bacon, institutionalized by Newton's bold statement that he did not make or require hypotheses. In inductive philosophies scientific reasoning characteristically goes from the ground up. We start with simple observations, make generalizations, test them, make grander generalizations. The end results are theories and laws of nature.

The other basic methodology is from the top down. It is the method of hypothesis. We make guesses, deduce testable consequences, conduct experiments, throw out the bad guesses that are refuted by experiment, and proceed to new conjectures. The method of hypothesis was a commonplace throughout the nineteenth century from the time of Whewell, and was made central, under the term "abduction," to the scientific philosophy of Charles Sanders Peirce. More recent is the hypothetico-deductive method, or Popper's methodology of conjectures and refutations. Larry Laudan (1981) has described the transition from one vision of science to the other, and argued that it is in the work of scientists, rather than philosophers, that we can understand why different generations emphasized different aspects of reasoning. That is not surprising, since the problems and practices of the sciences themselves changed over the years.

Let us start with an inductive account of the research on dolomite, and proceed to a hypothetico-deductive account. In the hands of philosophers this can be cast as an epic tale of a battle between inductivism and deductivism, Rudolf Carnap and Karl Popper slugging it out. In the life of the scientist the two visions of scientific practice are at worst binocular, and we learned how to use two eyes a long time ago.

The disciplinary label "philosophy of science" covers a manifold of practices and interests. There are philosophers who address specific problems of space and time, of quantum mechanics, or evolutionary biology, of brain science. Even when we turn to generalists, there are the metaphysicians and the epistemologists. Metaphysicians ask what there is, while epistemologists ask how we find out.

Even among the epistemologists there is something of a divide. Some workers want primarily to describe scientific activity and results. Others want to act as assessors and advisers, saying what programs and approaches are sensible, what reasoning is sound, how scientists ought to proceed. They aim at setting forth the right scientific method, or methods. This is not, nowadays, done by armchair reflection. Instead, one is supposed to look at the best examples of research and infer methodological maxims from success. Although this second type of epistemology is founded on close observation of scientific activity, its goals are normative. It is alive and well, as instanced by Larry Laudan's successive books (1977, 1996). Other philosophers of the sciences see themselves in a more modest light. There are particular instances where scientific work may be conceptually confused, and where a philosophical reflection may be helpful. But in general, say the modest ones, the people who know best what to do next in a science are the leading workers in the field, the scientists. Let us describe it, see how it works, comprehend what scientists do, but do not tell them what to do. Let us take them at their words, and not try to second-guess what they really find out. This approach is well exemplified by Arthur Fine's (1986) "natural ontological attitude" to the sciences.

I have no normative epistemology to advance, no general all-purpose normative stance whatsoever. There is no one scientific method; the sciences are as disunified in their methods as in their topics (Hacking 1996). I have from time to time criticized, often harshly, this or that bit of scientific research, and have argued that some competing statistical methods are more appropriate to some problems than others. Questions of method arise in context. These opinions in no way imply that one should ignore the great normative epistemologists. Popper thought that his method of conjectures and refutations is the fundamental and correct methodology of science, and that scientists ought to heed his advice. But we can make relatively nonjudgmental use of Popper, noting that the method he advocated is used and strikes us as sensible. What follows is ecumenical descriptive epistemology with hardly any normative implications.

Induction

According to the inductive method, we begin with observations. We connect observations so as to form generalizations. We check these out. We proceed by analogy to further generalizations. If we have a wide enough range of confirming examples, we think we have found something out, and add it to our store of probable and interesting truths, upon which we further build. For a more elegant statement, which comes from the end of the heyday of inductivism, I like to quote Humphry Davy. He did give general lectures on geology (1805/1980), but his inductive method was better stated in his introduction to chemistry.

> The foundations of chemical philosophy, are observation, experiment, and analogy. By observation, facts are distinctly and minutely impressed on the mind. By analogy, similar facts are connected. By experiment, new facts are discovered; and, in the progression of knowledge, observation, guided by analogy, leads to experiment, and analogy, confirmed by experiment, becomes scientific truth. (Davy 1812, 2)

Davy, like all others deeply engaged in their crafts, undoubtedly had very strong views about how a scientist ought to proceed. But notice that his account can be read in a modest and descriptive way: this is how we, as scientists, work. He continued with a charming example of globules of air on green vegetable filaments found in ponds and streams—an observation which "gives no information respecting the nature of the air." Then a wine glass is placed over such plants to collect the air. It is found that a taper burns more brilliantly in the collected air, than in ordinary air. "The question is put," whether this is true for all vegetables of this kind. After many diverse trials "a *general scientific truth* is established—That all Confervae [as he names this species of pond life] produce a species of air that supports flame to a superior degree" (1812, 3). In fact, his classification of pond life has long since been abandoned; Confervae constitute a spurious species. Yet Davy was not altogether wrong: just off the mark in summarizing an important fact about pond life, a fact that, as we would now put it, is essential for the ecology of the entire planet.

Arduino and Dolomieu distinctly impressed some facts on the mind by observation. We have all learned to say that observations are theory-loaded. Years ago I pointed out (1983) that a lot of relevant observations bear very little load of theory, and in particular, are not much laden with the theory to which they are taken to be relevant. A whiff of rotten eggs is theory-laden, yes, it is H_2S, but that is not a lot of theory to bear. The theory-loaded doctrines of Norwood Russell Hanson, Paul Feyerabend, and Thomas Kuhn taught us much, but need not be overdone. To see all observations as equally loaded with theory is in itself to practice theory-laden observation, that is, observation loaded with a theory derived from Hanson the philosopher.

When we move to the present, McKenzie and Vasconcelos are first of all guided by analogy. McKenzie suspects there is a simple explanation why no one found dolomite forming: no one was looking in sufficiently disagreeable places. Vasconcelos has a hostile environment near home in Brazil. By analogy, let us examine it. By experiment, new facts are discovered. Yes, upon culturing the swamp sludge, it is found to contain sulphate-reducing agents. A whole web of analogies comes into play, the analogy with Nadson's old work on a Russian salt lake. There is also the analogy with Folk's recent conjectures about nanobacteria. We are led to a new experiment. If we leave the bacteria in a medium for a year, do we precipitate dolomite? Yes, although we must test the precipitate in many ways.

Have the analogies of McKenzie and Vasconcelos become (to continue using the language of Humphry Davy) scientific truth? At most, only the first of the analogies just mentioned. There is a touch of rhetoric when our authors write that they have *conclusive evidence* that the bacterial production of dolomite can be achieved in low-temperature anoxic conditions in a relatively short time.

As for the more powerful analogies, about the origin of dolomite in world history, McKenzie and Vasconcelos are a long way from establishing a general scientific truth. And even if they are correct, and couch their discovery in terms of nanobacteria, their analogies might, like Davy's, be off the mark. Perhaps there is no general kind of living organism that we shall in the future classify as nanobacteria, no more than Confervae. And yet there might be a residual analogy that is as sound as Davy's basic guess at the oxygen-producing character of much vegetable pond life.

The Method of Hypothesis

Davy emphasized that most of us proceed through observations and analogy, forming generalizations that we put to experimental proof. Few of us are good observers.[4] Fewer still make lively analogies. And hardly any of us put analogies to the test. But Davy was surely right about one of the many types of valuable scientific mind.

There are other kinds of science. Some version of a hypothetico-deductive method has been prominent in the philosophy of science from Whewell through Popper. Popper emphasized that it does not matter much how you get your generalizations, so long as you can subject them to experimental test. The elementary logic of the method has been most simply explained in C. G. Hempel's introductory textbook (1965). We need clearly stated hypotheses from which we can derive test-implications, and we retain those hypotheses that survive tests. You can redescribe much of the work by McKenzie and Vasconcelos in just such terms. Popper favored bold conjectures. Take the conjectures about nanobacteria being among the earliest forms of life on earth. You cannot beat that for sheer chutzpah—until you move on to the extra-terrestrial.

Popper had strong normative views. He thought that the bolder a hypothesis, the more open to different types of testing and potential falsification, the better, the more scientific, the hypothesis. That idea was important, especially in the social sciences, at a time when there was a great thrust to inductive reasoning, to collection of data without any principle, and to generalization without enough hard questions being asked. One may see the harsher normative aspects of Popper's philosophy as having an important prophylactic function at a certain point in the evolution of the social sciences. But now almost everyone in the social sciences has learned at least to mouth the injunction that hypotheses must be testable. After the Popperian dust has settled, we see in plain scientific work, such as the study of dolomite, a happy mix of both induction and analogy (as described by Davy and many others) and conjecture and refutation (as prescribed by Popper).

SCIENCE STUDIES

Recent science studies make us notice other aspects of scientific work. In the examples that follow, my aim is not to defend either science studies or the old-time philosophy, but to save us from what William Blake called "single vision."

Edinburgh: Interests and Symmetry

The Edinburgh school of sociology of science began by emphasizing the ways in which interests play a role in the sciences. Certainly that is true for dolomite. Magnesium replaces calcium in limestone. So what? Magnesium carbonate is less voluminous than the calcium carbonate that it replaces: 12 percent less, to be exact. So dolomite is porous. Nature abhors a pore, or rather, delights in filling them. From the beginning, dolomite fascinated prospectors as a potential location for valuable minerals that had been pressed into the tiny holes. Now the interest is oil. Much of the known petroleum in the world is trapped below layers of relatively impervious limestone. Dolomite could be the perfect way for nature to store her oil: all those lovely pores. The petroleum corporations have not been idle: a great deal is known about dolomite.

Members of the Edinburgh school were often thought to imply that interests affected the actual content of a science. I am not sure about the extent to which this accusation is justified. Interests have a lot to do with the questions that are asked, with the direction of research, and the resultant form, as opposed to the content, of the science. The enormous commercial importance of dolomite, as a container and cap for petroleum, has had an obvious effect on how questions about dolomite are answered, given that the questions are asked. But this is more a matter of the form of the knowledge than of its specific content, what we find out once the questions are asked.

The Edinburgh school is also famous for its strong thesis of symmetry, briefly discussed in Chapter 3. The idea is that an explanation of why a group of investigators holds true beliefs should have a very similar structure to an explanation of why another group holds false beliefs. The early days of dolomite serve us well to illustrate this doctrine. There is what we now take to be the correct account, furnished by Arduino, and the incorrect account furnished by Saussure and accepted by Dolomieu. Arduino thought he had a magnesium compound, while Saussure thought it was an aluminum compound, and claimed to prove this by chemical analysis. The two cases seem symmetric. Arduino did not reach his belief *because* it was true (in Chapter 3 I inveighed, on logical and linguistic grounds, against saying anything of the sort). The explanation of why Arduino reached his correct conclusions will be of very much the same sort as the explanation of why Saussure reached his mistaken conclusions.

Networks

Bruno Latour and his colleagues advance a different way to study science. In *Science in Action* (1986) he lays great emphasis on the building up of scientific networks, both for the growth of scientific knowledge and for its stability. The better you are networked, the more likely it is that your beliefs will catch on and persist. Dolomieu was superbly networked as Napoleon's scientist, and also because he was located in what was then the finest center for mineralogy in the world, the Ecole des Mines.[5] That does not explain why Saussure and Dolomieu thought that dolomite was an aluminum compound, but it does help explain why their doctrine was widely accepted. Arduino was not networked.

Skeptics of the importance of networking will rightly point out that Arduino's account triumphed in the end. We might once again invoke Lakatos's idea of a research program. The Arduino research program of magnesia composition and magnesium ions in sea water replacing calcium was, over the years, both theoretically and empirically progressive. It retained a hard core of beliefs (magnesium, sea water) but regularly opened up new research directions, despite "wallowing in a sea of anomalies" (Lakatos 1970).

We can see McKenzie and Vasconcelos as participating in one of the current thrusts of the Arduino research program. Yet we should also see that this program persists in virtue of networks of knowledge, instrumentation, publication. McKenzie has access to the Idaho National Engineering Laboratories (hardly one of your trendier sites for modern science). The group makes use of all those black boxes, to use Latour's admirable description: the scanning electron microscope, energy dispersive X-ray spectra, X-ray diffraction, a transmission electron microscope of old vintage. These are now the standard black-boxed equipment of a materials science laboratory. The ETH group was not acquainted with the scanning tunneling microscope or the atomic force microscope, even though those devices were invented almost next door.[6] This is a nice illustration of the way in which what Peter Galison (1986) calls instrumental traditions help direct the course of research.

Notice how McKenzie and Vasconcelos extended their network back in time. McKenzie wrote a survey article on the dolomite problem. She introduced a number of figures, such as von Morlot (writing in 1847), who seem to have disappeared from earlier twentieth-century accounts of the problem. He became an ally in recasting how we think about the problem, an ally against certain more established sedimentologists. Her revival of the forgotten Arduino against the admired Dolomieu is part of a strategy for entrenching her—and his—program. She finds a particularly useful dead ally, Nadson, with his Russian salt lake and his sulphur-eating bacteria of 1924.

McKenzie has a great potential ally in Folk, who is promoting nanobacteria. One thing that main-line bacteriologists do not want to hear about is sedimentary petrology! It will take quite a few voices to get them to listen. Next step, convince someone to do some DNA sequencing of the bugs in the mud of Lagoa Vermelha. Already there is a microbiologist from Dübendorf, Switzerland, on the author list of the Letter to *Nature* (Vasconcelos et al. 1995). It is absolutely essential for the success of the program to involve a whole range of fellow disciplines.

There is something classic about the dolomite story. When thinkers—from Dr. Johnson (against the immaterialist Bishop Berkeley) to Steven Weinberg (against cultural relativists)—want to say that something is real, they resort to rocks. Dr. Johnson kicks one. Steven Weinberg compares the reality of Maxwell's Equations to the reality of rocks. So I have told a story about what some distinguished thinkers seem to regard as the most unquestionable reality, namely rocks. You want rocks? We have rocks.

On the other hand, when Bruno Latour wishes an example of something nonhuman collaborating in the work of science in action, he has many examples, but his long-time favorites are bacteria. Pasteur's knowledge arose only when he had got the bacteria to work with him. That story is brilliantly told in *Les Microbes* (translated as Latour 1988).

You want microbes? We have microbes, microbes smaller than anything imagined by Pasteur, nanobacteria. And they are working for us. We can smell the results of their labor, when we uncap the vial that has been sitting in the refrigerator for a year, it's that smell of rotten eggs, which means so much to the investigator. "They're there, those tiny bugs, hooray!" (We can imagine Vasconcelos exclaiming that, in Portuguese.) The allies, which we extracted in a plastic pipe from a swamp in Brazil and stored in a flask in Zurich, are performing their expected tasks, reducing sulphate ions, and thus making dolomitization possible.

Why should one resist this version of events? I do resist the suggestion that *nanobacterium* is an interactive kind, in the sense of the earlier chapters in this book. It is true that "ally" is too interactive a metaphor for total comfort. But from other perspectives, it is a positively attractive metaphor, to be resisted only if a monomaniac tries to turn it into a monocular vision of scientific activity.

THE THREE STICKING POINTS

The sticking points were claimed as just that, points at which people stick. Rocks are more down to earth than Maxwell's Equations, but the sticking points work in pretty much the same way for dolomite as for electrodynamics. Rather than attempt a general survey, I shall briefly sketch why my own score, on each of the three sticking points, is about the same for dolomite as the scores announced at the end of Chapter 3. Then I invite you to undertake the same exercise, and compare your scores for dolomite with the scores you had at the end of Chapter 3.

Contingency. I think that once the question of the chemical composition of dolomite is asked, then an answer in terms of magnesium is pretty inevitable. But I do not see that this way of thinking about substances on the surface and in the crust of the earth is inevitable. I do not mean only that most of us who have walked in the Tyrolean Alps have never even noticed the dolomite, or could notice it if we tried without further instruction. Dolomite is a sorting that has assumed prominence precisely because of human interests, in particular, an interest in petroleum. A wholly different set of questions could have arisen, provoking a thoroughly nonequivalent geology and geophysics. My score on contingency is only 2 out of 5 precisely because I hold that once the questions are asked, the answers are predetermined. Different readers will have different scores, reflecting fundamentally different attitudes to the nature of scientific research.

Nominalism. This issue is often put in terms of "natural kinds." Does the world of itself have an inherent structure of which dolomite is a part? Dolomite is a very healthy example here, because many philosophers of natural kinds would not think that it is a natural kind at all. They want the natural kinds to be cosmic, to be like the elements and the fundamental particles, and not to be messy mixtures formed by mysterious processes. There is something of a continuum between sedimentary rocks that are primarily magnesium carbonates, and those that are primarily calcium carbonates. Dolomite is a peculiarly human sorting, one that would have little significance, in the cosmic scale of things, had it not assumed such an important role in the search first for minerals and then for oil. Hence many natural kind philosophers ought to be inclined to be quite nominalistic about dolomite. For my part, I score myself 4 out of 5, just as at the end of Chapter 3.

Stability. There is not much stability to explain. The history of the dolomite problem is as much a matter of myths and their debunking, as of steady growth of stable knowledge. When it comes to hypotheses about the formation of the large masses of dolomite in the earth's crust, there is no stability to explain. Nobody knows for sure. People are betting with their careers, with their lives, on this or that hypothesis, but no one, in all honesty, knows what will pan out even in a decade. Some beliefs have stabilized; for example, that there are large masses of porous rock that is primarily magnesium carbonate. Are the reasons for the stability of this belief external to the science, or internal? That is the way I left such questions at the end of Chapter 3. I am inclined to say that there are internal explanations for the stability of the belief in magnesium composition. The explanation has nothing to do with stable networks or human interests. But that answer may be misleading. This is because when belief, in the words of Humphry Davy, "becomes scientific truth," then all the grounds for the belief become internal to the science in which it is located. I shall not pursue the point. The dolomite problem leaves philosophical questions of stability untouched, precisely because it is still a problem.

Chapter Eight

THE END OF CAPTAIN COOK

In 1994 the eminent anthropologist Marshall Sahlins published *How "Natives" Think. About Captain Cook, for Example,* a retort to the work of another distinguished anthropologist, Gananath Obeyeskere's 1992 *The Apotheosis of Captain Cook: European Mythmaking in the Pacific.* I discussed Sahlins's book in *The London Review of Books* in September, 1995. My review did not mention social construction but spoke of the culture wars. Oppression, postcolonial history, multiculturalism, and many more topics had been thrown into the ring by this controversy, which became emblematic of a lot of academic wars going on in America today. In fact soon afterward yet another notable anthropologist, Clifford Geertz (1996), wrote the debate up in *The New York Review of Books;* on the cover his review was announced under the simple headline "The Culture Wars."

Since social-construction polemics form one aspect of the culture wars, my editor, Lindsay Waters, thought that this piece would nicely round out the discussion, a bit of icing on the cake. As with chapters 5 and 6, I have not attempted to update this material. Any updating would be out of date next month. *Current Anthropology* for 1997 (volume 38) continues the argument, with entries from both main protagonists. A philosopher-ethnographer has just put his hat in the ring (Bravo 1998). Obeyeskere (1997) added a 63-page reply to Sahlins to the second edition of his book. He called it "On de-Sahlinization," which was in fact Sahlins's own pun, the product of his relentless pun-machine which marches through his book.

There are less obvious plays on words. *How Natives Think* was the English translation (1912) of the title of a classic work by French anthropologist Lucien Lévy-Bruhl. The original French was *Les Fonctions mentales dans les sociétés inférieures* (1910). Other book titles of his were *L'âme primitive* and *La Mentalité primitive.* That is the picture: the primitive/inferior mind/soul/thought/society. Turning Lévy-Bruhl's title word *Think* into Sahlins's *Think About* is a wonderful trick for undoing the whole rhetoric of primitive and inferior minds. It also switches us away from highbrow philosophical language ("think" as one of the profound things that people do, I think therefore I am, Descartes, the whole history of modern philosophy). It returns us to ordinary intelligible English (thinking about so and so is one of the ordinary things that ordinary people do). But there is more to it than that, because Sahlins thinks that Obeyeskere, spokesman for the rationality of Hawaiians, is denying those self-same Hawaiians the right to speak for themselves. Obeyeskere, claims Sahlins, uses the very same presuppositions that led Lévy-Bruhl to call natives primitive. The presuppositions are of course applied with the opposite intention. Whereas Lévy-Bruhl's natives were primitive, Obeyeskere's natives are thoroughly rational. Here we have the odd spectacle of the "relativist," Sahlins, saying that the imperial explorers had one right version of the world (to use Nelson Goodman's language). In contrast, the rationalist, Obeyeskere, denies that the inventors of the very idea of reason had a right version, and gives to the "natives" a rather Western rationality. Perhaps it is the paradoxical character of the debate that has made it so enduring.

For all the heat about the apotheosis of Captain Cook, I would like to think that *How "Natives" Think* will lead its readers to Sahlins's earlier books mentioned below, and also to Obeyesekere's contributions to studies of South-East Asian cultures. But for someone who just likes stories, the best treat of all would be J. C. Beaglehole's (1968) edition of the journals of Cook's three voyages. The quoted descriptions of Cook's voyages used in this chapter are all taken from Beaglehole. Some readers, including Obeyeskere, may take issue with Beaglehole as a historian of the ocean and of the period, but his edition of Cook and Cook's fellow seamen is a marvellous read.

I write as a complete outsider. I do not think I read a careful word about the South Seas after a nautically minded uncle gave me the *Mutiny on the Bounty* trilogy when I was 11. Since then there has arisen an entire ethnographic industry whose laborers are called "oceanists." These experts may take issue with the details of assertions made by either Obeyesekere or Sahlins or both, details that an outsider will not even notice. A caveat, though: I suspect that all these adventures, including the confrontation between Sahlins and Obeyesekere itself, will appeal more to the grown-up boys and girls who wallowed in *Treasure Island* at the age of seven (in my case, a birthday present from the same nautical uncle) than to the grown-ups who did not. As I said in Chapter 1 about the Sokal dispute and the science wars, we are on the edge of a spectator sport.

My mention of the classic tales of adventure in the South Seas is not incidental. Apotheosis means deification. The issue throughout Sahlins's book is whether, and at what point, Cook was deemed or made a god in the ample Hawaiian pantheon. Obeyeskere is equally interested in the European pantheon, and the way in which Cook became the stuff of legend. Where Sahlins argues that Cook fitted almost seamlessly, at least at first, into Hawaiian myth, Obeyeskere argues that he was the incarnation of a European myth model. "One of the most enduring ideas in Western culture is that of the redoubtable person coming from Europe to a savage land, a harbinger of civilization who remains immune to savage ways, maintaining his integrity and identity" (Obeyeskere 1992a, 11). That's Shakespeare's Prospero. But there is also Conrad's Kurtz, "the civilizer who loses his identity and goes native and becomes the very savage that he despises." The European apotheosis of Captain Cook took place within these visions of the archetypal hero (or anti-hero) on the very edge (or over the edge) of the Western world. Those myth models persist to this very day, John Glenn as Prospero, perhaps, or Lieutenant William Calley at My Lai as Kurtz.

THE OPPONENTS

Marshall Sahlins is an anthropologist whose theoretical work is renowned, and whose expertise as an "oceanist"—a student of the island civilizations of the Pacific, past and present—is beyond compare. His book is a splendid work of refutation and revenge, judicious but remorseless, urbane yet gritty. It is an adventure story in itself, and a stepping-stone to better ones. My only regret is that this book—one of a quartet—may be more widely read than Sahlins's *Historical Metaphors and Mythical Realities* (1981) and *Islands of History* (1985). There is also the lavishly illustrated two-volume *Anahulu: An Anthropology of History in the Kingdom of Hawaii* (Kirch and Sahlins 1992), an extraordinary collaboration between Sahlins and an archaeologist, creating an account, seen from both inside and outside, of an isolated society under cultural siege, and of the ways that it responded to the ensuing changes.

How "Natives" Think is focused on the narrow question of whether the Hawaiians, on their first prolonged encounter with Europeans, regarded the white men as superior human beings, and even took Captain Cook to be their own god Lono. Obeyeskere thinks not. He argues that the story is a European myth foisted on Hawaiian self-memory by British and other foreign chroniclers. His *Apotheosis* is an angry and powerful attack on what Sahlins had written in his earlier books. *How "Natives" Think* is Sahlins's response to Obeyeskere's salvo.

This impassioned debate about "what actually happened" appears deceptively simple. Either (1) Hawaiians recognized Cook as a god, almost on arrival (Sahlins), or (2) they were not plain stupid and, after killing Cook, found it politically convenient, given local power struggles, to deify him (Obeyesekere). The dialectic of the confrontation between the two historical ethnographers makes the apotheosis the turning point: Obeyeskere writes that Sahlins got it wrong in his previous work, and Sahlins reacts, angrily. The choice of answers, (1) or (2), is not central to Sahlins's earlier books. It is critical only to chapter 4 of *Islands.* Most of Sahlins's work is in the tradition of Lévi-Strauss. Instead of the doctrinaire structuralism of some of Lévi-Strauss's followers, who take structures of autochthonous societies to be instantaneous, outside of all time, Sahlins studies how they change in time. He is especially interested in how the conceptual frameworks and practices of one people evolve after a first encounter with another people, especially with the arrival of a colonial power. The title *Historical Metaphors and Mythical Realities* is just right. Sahlins argues in that book that we ought not to make an issue of the differences between Hawaiian myths and historical reality. Cook's arrival and subsequent events, he says, jibe perfectly with prior Hawaiian beliefs, even if the English vision of what was happening looks very different from the Hawaiian one. What scholars call the historical events served as a metaphor for Hawaiian expectations; what scholars think of as the reality was decoded by Hawaiians in terms of their established beliefs, the very beliefs that Europeans called myths.

Sahlins uses his analysis to explain amazing goings-on during Cook's presence at the islands, and subsequent transformations in Hawaiian culture. In 1993, in the *Journal of Modern History,* Sahlins extended his argument: the glory days of ethnographic field work with an "unspoiled" people are over. They were mostly fantasy anyway, a dream of the purity of primitive man in isolation. Everybody is contaminated by the capitalist world order, which Sahlins calls the World System. In fact people have always interacted with strangers, and adapting to the world system is new only in that the system seems (but only seems) to be homogenizing. Sahlins also finds subsequent cultural revivals, whether resisting or collaborating with the world system, anthropologically interesting. They are not just phoney. Sahlins insists on this no matter how much what is "revived" has been transformed. Not even the hula-dancing flower girls who greet tourists at the airport in Honolulu are to be dismissed from historical ethnography. The task is to understand precisely how an earlier web of ideas and practices adapts, internalizes, exploits, or re-sees the interaction with what was once alien and more powerful.

Right or wrong, Sahlins's ideas are deeply challenging. They make the question of the apotheosis of Captain Cook relatively small potatoes. Not small potatoes for the culture wars, however. Enter a brown man, Obeyesekere, who is a professor at Princeton but who grew up on a colonized island, Sri Lanka. He says that a white man, a professor at Chicago, is foisting white myths onto islanders and perpetuating the fantasy that natives first see Europeans as gods. That may be a delightful thought for the white man, but it quite ignores the good sense of the natives. If the Hawaiians ever deified Cook, Obeyesekere argues, they did so after he was dead, and then only for rational, pragmatic, and intelligible political reasons. Sahlins, holier than Obeyesekere, retorts that Obeyesekere is the imperialist. By treating Hawaiians as political players not so far off from rational choice theory, the Sri Lankan denies the islanders "their own voice" (as one used to say in connection with gender). We have the bizarre spectacle of a Buddhist working in the American academy using American pragmatism to silence Polynesian culture forever.

This is old-fashioned pamphleteering. It helps, in reading our two polemicists, to have a taste for the modes of arguments of the Augustan Age in England. We have a nice example of Sahlins's thesis about one group appropriating something from another culture. Obeyeskere and Sahlins, themselves modified and adapted by the historical events of a couple of centuries, have adopted the tools of Addison or Swift, just as those authors acted out what they saw of Rome through eighteenth-century culture.

It is a very nice innovation, in the culture wars, to have arguments—

powerful, cogent old-fashioned arguments, premises, evidence, deductions, inductions, qualified probabilities, hypotheses, tests, refutation—on both sides. It is good to be reminded that arguments work. I entered this fray with a bias toward Obeyesekere's thesis. Like him, I have a set of fairly strong "Enlightenment" universalist prejudices. Humans, everywhere, are what I call reasonable, even if every culture also harbors its dangerous lunatics. I left the debate with the conviction that despite retaining a preference for the principles that govern my prejudices, and despite a lot of side issues on which I align with Obeyesekere, Sahlins is right about the so-called apotheosis of Captain Cook. Or, strictly speaking, that is where I think the balance of probability lies, based on the evidence presented.

THE DATA

There are two sets of data, British and Hawaiian. The British data are written down and numerous. Cook's previous voyages had been sensational, the toast of Europe and the American colonies. These expeditions, equipped with supernumeraries of artists, astronomers, and Kew gardeners, brought back the first vision of the Pacific. Bernard Smith, the Australian historian of art and exploration, argues that these voyages were the first to flood Europe with images of new lands, new peoples, new worlds. Perhaps thanks to the fashionable enthusiasm at the time for anything to do with Cook, many narratives of the third voyage, including cameo reports of the captain's death, have been preserved. Cook's own journal is meticulous, faithful to its task as the commander's record, and an unusual number of his men kept journals.

And here is something to strike fear into the heart of every student of evidence. We know an enormous amount about the voyage and quite a lot about the captain's death, and yet precious little. The records are written by various hands, although Cook's most famous officers, Midshipman Vancouver and Captain Bligh, were fairly reticent, as if awaiting their own glory or catastrophe. The stories hang together well, except for that of Corporal John Ledyard, a marine of Groton, Connecticut. (Thomas Jefferson encouraged this legendary adventurer to walk from Siberia to Nootka Sound, off Vancouver Island, and on to Virginia. He was arrested at Irkutsk and returned to the Polish border; later he was hired to help explore the Niger, but died en route, in Cairo.) Ledyard was the chief journalist aboard the ships—yes, each ship ran a weekly newspaper, of which no sheet survives. All authorities but Obeyesekere distrust Ledyard—more on that anon.

Here is what we do know about Cook's voyage. The two ships had a complement of 112 men, plus supernumeraries, who were replaced or traded from time to time. Only 46 were seamen and servants. There were a great many landsmen: draughtsmen, surgeons, cooks, carpenters, sail-makers, smiths. The chief armorer had to be sent home from the Cape of Good Hope as he had spent the pleasant days on the trip south minting false coin. It was standard for many of the seamen to be (literally) jacks-of-all-trades. This was a little English society, right down to the counterfeiter. In many ways it was a microcosm of the English village, even down to the flora and fauna: a horse and a mare for breeding, and "as many Sheep Goats Hogs Rabbits Turkeys Geese Ducks, a peacock and Hen, as they could conveniently make room for" on a 462-tonner, 111 feet long. The animals were brought with the somewhat zany but philanthropic thought that if these breeds naturalized on Pacific islands they would much improve the lot of the inhabitants; food-plants were carried, with the even zanier thought that Tahiti could profit from a spot of English gardening. There is a feeling of knowing everything and nothing about the sociology of this society. The hermeneutic rule would be: if you want to know what it was like among these men, use as a partial model long voyages now or in wartime. Think of the lives of men on one of those endless tours of duty in a nuclear submarine, but add in a thicker layer of drink, indiscipline, desertion, cruelty, and sex.

Despite all the nastiness, *Treasure Island* is a pale pastiche of adventure compared to the wonderful journals of our navigators and their crews. Take the "Private Signals" for the final voyage, which lasted four years, for use in the event that Cook's two ships were parted and met again. There were complicated signals if they met on the horizon to guard against privateers and enemies. When they were within hailing distance (which means you could recognize a man you'd been sailing with for years, at least through a spyglass): "he who hails first shall ask, *What ship's that?* then he that is hailed shall answer *King George* then he who hailed first shall answer *Queen Charlotte,* and the other shall answer *God Preserve.*" If the crews really got out of touch they were to leave messages in bottles at preassigned beaches or map readings.

On the Hawaiian side things are murkier: no written records, only memories, tales, and songs, usually recorded by missionaries or island converts. And it is here, in this evidential vacuum, that Obeyesekere and Sahlins diverge on critical issues, such as how and why Cook died, and whether he was esteemed a god at first meeting. Obeyesekere thinks that the missionaries and their converts were wedded to the European god-myth story, which was then internalized into Hawaiian legend. In his view, the most one can get out of the missionary stories is a subtext. Sahlins, in contrast, mines these tales for clues of practices from earlier times. For example, Obeyesekere claims that Cook's bones were given a deification ritual appropriate for a dead chief. Sahlins argues that we can tell clearly from Hawaiian accounts that there were two altogether distinct sets of things that can be done with bones. There is one practice for deification, and another for dealing with a dead god-king. The British records and the Hawaiian stories indicate, with unusual unanimity, that Cook's bones got the latter treatment. Sahlins does not use the Hawaiian texts to argue that the British accounts must be right, on the grounds that both sources pretty much agree. That is not what interests him. Instead, he uses the Hawaiian texts to infer new ethnographic information with which to interpret historical records that hitherto did not make much sense.

THE VOYAGE

I had better say a little more about what happened on the voyage. It is a story so familiar to oceanists that they will tune out; for the rest of us, however, it begins (although itself *in media res*) on 6 July 1776, just two days after the American Declaration of Independence. The secret instructions from the Admiralty to Captain James Cook, Commander of His Majesty's Sloop the *Resolution,* open: "Whereas the Earl of Sandwich has signified to us His Majesty's Pleasure that an attempt should be made to find out a Northern passage by sea from the Pacific to the Atlantic Ocean . . ." Cook, aged 48, finally in possession of a sinecure, was sent out in command of the *Resolution* and *Discovery* to find the North-West Passage across the Arctic, in the opposite direction to the more familiar searches for the impossible route. He encountered the Hawaiian Islands on the way.

The only remarkable thing about the first discovery of the islands is that it appears that no European had successfully sailed north from the South Seas. Remember that all this happened two centuries ago: for two full centuries and more before *that,* there had been a regular shipping service between Lima and Manila, and the South Seas were bustling with privateers.

When Cook arrived, Hawaiians wanted iron. They had quite a few tools using iron, obtained, it is thought, not by trade with more southerly islands, but taken from the driftwood of sunken ships. The skeptic in me wonders, is it really true that these people had no previous contact with Europeans or their artifacts?

Cook sighted Hawaii, circumnavigated it, was welcomed as nowhere else by islanders, went north, encountered a 12-foot wall of ice in midsummer at almost the northern extremity of Alaska, came back, circled the island three times, was met again with joy, and left. Cook's ship, the *Resolution,* had been fitted out privately for his second voyage, and was magnificent. For the third, it had been refitted in Deptford Naval Yards, a place of scandalous corruption and patronage. It was leaking before it left English waters. The rigging of its companion ship, the *Discovery,* was properly fitted, and was almost as good four years later as when she set out; *Resolution*'s failed over and over again. (True to form, a naval official stated that Cook, the greatest navigator of the age, knew nothing about rigging.) Hence masts failed, one after the final departure from Hawaii. The ships returned, to be treated to thievery and hostility by the islanders. In the end, the *Resolution's* cutter—her largest boat, and irreplaceable—was stolen. Cook reacted in what everyone agrees was an uncool way, landing with firearms, and shouting in panic. He and four marines were killed in the ensuing melee.

The goings-on, during Cook's visits, can only strike one as, well, strange. Problem: why if the ships were welcomed with so much good faith and no thievery beyond help-yourself on both the first and second trips around the island, did they return after saying their last farewells to hostility and disaster? Many other curiosities. Why, for instance, did the island women seem so eager for sex with Europeans? Even Obeyesekere takes this as a brute fact, not the exaggerated recollections of European machismo. Cook at first tried to prevent sexual encounters. He did not want to transmit disease to the locals, but he eventually gave up, or gave in. One has the feeling that the ship was swarming with women. That presented difficulties of various sorts. Sailors in their bunks would, it seems, pull nails out even of the hull to give as presents to their ladyfriends; at the same time the island lads in canoes were pulling nails out for themselves.

Violence

Why, when Cook was increasingly violent in dealing with annoyances, treating offending Hawaiians with cruelty and shooting at others, were relations so wonderful, till the end? Obeyesekere's basic answer to the sea-shift in Hawaiian attitudes is simple and plausible: Cook was falling to pieces throughout the third voyage. He was erratic, irritable, forgetful, unpredictable, given to bursts of anger, violence, and cruelty. Formerly the perfect navigator, he reckoned wrongly, omitted to survey what he ought to have, and unaccountably circled Hawaii three times on his return. He mismanaged his crew. Captain Bligh learned his lessons all too well from Cook. Flogging was the rule for those on board who made mistakes, and for Polynesians who made trouble. Long before reaching Hawaii, offending Polynesians who had been taken on board from other islands to the South were receiving three or four times more lashes than the maximum of twelve a day prescribed for seamen. Readers more familiar with Herman Melville than Captain Cook may recall that *Billy Budd* is set in 1797, five years after Cook was killed.

Cook became obsessive about his conviction that fresh food prevented scurvy. On the second voyage his complement of 118 men lost only one due to sickness in three years (health statistics tell us that several more would have died of disease had they stayed at home). That was a triumph, not an easy one. On the third voyage he managed to force sauerkraut on his men—however, a beer made from sitka spruce needles was not only revolting (and fairly poisonous, I should think) but also incited revolt. Off Hawaii he made a concoction of sugar cane which was to replace grog. He even tried to starve the seamen into submission. The ship had become the site of a culture of violence. Correspondingly, the slightest annoyance from islanders was treated with increasing violence and quite untypical misjudgment, hasty reaction, brutal punishment. Guess what? The Hawaiians, having greeted the visitors with good will, having got all the iron they could hope for, said good riddance. When an angry Cook returned, furious at everyone for bad rigging and a broken mast, the islanders went on a stealing spree that culminated in the theft of the ship's cutter. About the time Cook went ashore, a Hawaiian had just been killed by sailors at the other end of the bay. This was hardly the first victim, although Cook was not responsible for the very first Hawaiian fired at with ball (rather than pellets of lead shot) and killed by his men. Not too surprisingly, when Cook and the marines met mild force with shot—panicking when surrounded by thousands of angry people—a scuffle broke out. Cook and four of his marines were killed. Obeyesekere skilfully weaves a lot of circumstantial historical detail into what I have just sketched, but I hope I have conveyed the tone of his analysis.

Was it like that? The judicious Beaglehole never suggests a reign of terror, but he does make plain that Cook repeatedly changed the routine, made many errors of judgment, had outbursts of rage and apparent lapses of memory during the third voyage. The captain has previously been diagnosed by scholars with this or that flavor-of-the-day physical/mental disorder. Today he would be a candidate for early-onset Alzheimer's, except that he doesn't seem to have the pedigree.[1] But despite attempted desertions—rational acts in paradise—and occasional near-mutinies, Cook does seem to have kept the affections of most of his men—that is, except the American corporal, Ledyard. Obeyesekere trusts that adventurer more than anyone else in this business. We too should pay attention to a libertarian anti-English amateur journalist of those days.

I should guard against the thought that the English were violent and the Hawaiians lovely and lovable (by our standards), peaceable, singing, dancing, happy, adoring of lovely women, politically correct, ecologically minded, and sexually liberated—the sort of stuff that the Hawaiian tourist industry promotes. By our present lights, they were not such sweetie-pies two hundred years ago. The traditional Hawaiian year is divided into two parts, a point essential to Sahlins's analysis. During one month there is a period of renewal, war is forbidden, and affairs are largely in the charge of priests who honor the mythical return of a god and ex-king Lono. Then, they made love, not war. For the other eleven months, however, war was almost a norm. There was certainly a war in progress before Cook arrived, though in abeyance because of the month off. This month was characterized by its own kind of violence. Human sacrifice was an integral part of the renewal ritual. The return to power of the eleven-month king seemed to involve his eating two eyes, one of a bonito tuna and one of a sacrificed human. Missionary reports are consistent and unequivocal that throughout the entire year a great many infants, mostly girls, were routinely killed seconds after birth (and only then). The disproportion of live males to live females was said to be so great that I can hardly credit it. (More grist for the deMause theme of Chapter 5, about the universality of child abuse.) Why did they murder the baby girls? The chief end in life is to be a warrior. Warriors live short lives because they get killed early. Hence a society needs a lot more boys than girls.

Hawaiians, then, had what we, if not they, might think of as a culture of violence. Perhaps I am supposed to condemn the British culture and not the Hawaiian one, on (1) the excellent Millian ground that the Hawaiians left other people alone and did not seek them out, while the British did. And (2) the dubious suggestion that I am descended from the British culture and may condemn my own lot but should not criticize others. I am in favor of moral parallelism here. We are dealing with two great seafaring peoples, each possessed of a formidable nautical technology, and each given to a lot of violence.

God

Formally, Obeyesekere and Sahlins confront each other only when it comes to god. There is nothing seriously incompatible between Obeyesekere's vision of an English culture of violence, and Sahlins's ethnography. Ironically, we can read Obeyesekere as, among other things, the ethnographer of a fragment of an English village: a society of young males, out to sea for years, a society about which Sahlins is discreetly silent. Or almost. He has 17 appendices rebutting particular bits of Obeyesekere's version. One is entitled "On the Wrath of Captain Cook." This is, as throughout Sahlins's polemic, first-class on critical details of Obeyesekere's account of why Cook was especially angry at Hawaiians, and at the world, during his return to the island. "Obeyesekere's speculations on Cook's wrath give rise to an elaborate set of makeshift interpretations whose truth values range from the historically unknown to the ethnographically unwarranted, passing by way of the textually unproven." That is a fair sample of Sahlins's icy anger. It opts out of Obeyesekere's more global claim—truly marred by some dubious Hawaiian ethnography that I have omitted—that Cook was falling apart and the Hawaiians had good reason to lose their tempers.

I have managed to put off the god issue long enough, partly because, as I said at the start, it is small potatoes until we start reading it through our own present-day myths of culture and oppression. Cook arrived at the time of the winter festival, when legend had it that the god Lono returned from a distant land over the horizon. Usually he did it every year, in symbolic form, but this time it was, amazingly, for real. During the festival the center of celebration toured around the island. Cook circumnavigated in just the right direction at just the right time. So Cook is greeted not only as a superior being, but literally as a god. But we're not to understand this to be like God, utterly absent from anything earthly, or like Christ, made incarnate by divine order. Nature and supernature mingle happily in Hawaiian culture. And we are not to imagine everyone thought Europeans superior or those at the top divine. Sahlins invokes what the philosopher Hilary Putnam calls the division of linguistic labor: just as we use experts to tell gold from nongold, priests tell god from nongod. The priestly hierarchy was very different from the kingly hierarchy—a fact too little studied by Obeyesekere—and it was the influential one during the winter festival. Cook and his men were welcomed as gods. Why then a later killing? Because when *Resolution* returned with a sprung mast in dire need of repair, the winter festival had ended. It was now the time of the eleven-month king, who in the normal course of events was ritually restored by a sacrifice of a representative of the loser king, once Lono, who then went out beyond the horizon for 11 months. The islanders were confused when Cook came back, but rationalized the possibly accidental effects of a scuffle by performing the appropriate rituals, including peeling off the flesh, burning the surfaces of the bones and distributing them—each rank in the hierarchy received its allotted bone. There is an immense amount of detail here, and whether or not it is fitted rightly, Sahlins does a masterly job of producing a coherent narrative, from within a conjectured Hawaiian structural space of ideas and practices.

To which Obeyesekere protests that the whole god business is a European myth. The question is whether it fits Hawaiian myth. Obeyesekere argues that the Hawaiians were sensible people. Some entertained the possibility that Cook might be a god, but after examination this was rejected. They did greet him as a chief. The early rituals in Cook's encounters had to do with that status. After his death, those who owned the bones found it expedient, for political reasons, to deify this powerful intruder. A simple point: god after death, but not before. Obeyesekere argues at length for his hypothesis, which clearly rides in tandem with the culture-of-violence analysis of Cook's command.

UNIVERSALISM

I said I had an initial prejudice in favor of Obeyesekere. Perhaps I first extrapolated it from some maxims of David Hume to the effect that if you want to know why men did such and such in Rome, look for models among the Edinburgh politicians of the day; conversely, if you want to predict how your city council will react, consider how some others, Romans say, did in comparable circumstances. That is a version of universalism: expect human nature to be pretty much the same any place, any time. Noam Chomsky is our most famous universalist today. The most remembered universalist from the beginning of the twentieth century is Sigmund Freud, who wrote about human nature, the nature of humans anywhere. Obeyesekere is a universalist, though what he holds to be universal is closer to Freud than to Chomsky. Despite an avowed interest in psychoanalysis, he cannot deploy Freud because we know so little about the childhood of any of the actors, whether Cook and his men on the one side, or the Hawaiian king, priests, or people on the other. In an essay on cannibalism among the Maori, another universalism, that of Jung and his archetypes, makes a striking appearance (Obeyesekere 1992b).

Sahlins pointedly presents universalists as being sucked into a naive empiricism—everybody sees the world in the same way. So, he implies of Obeyesekere, the Hawaiians perceived a *man,* Cook, a sense datum. And they acted out of, as Obeyesekere himself says, "pragmatic rationality," a kind of rationality that established itself in Europe at the time of the great empiricists. Thus Obeyesekere is saddled with wicked empiricism and bourgeois ideology. Sahlins repeatedly argues that it is Obeyesekere who is the imperialist, who denies the Hawaiians their own historical voice. Good polemics, but it does not probe the depth of universalism (which underlies those shallow tags of rationalism and empiricism). Obeyesekere's real difficulty is in fitting his universalism into a careful survey of the ethnographic data. There is the matter I mentioned earlier, of whether what happened to Cook's bones was deification, or the treatment accorded at the end of the winter festival to the sacrificed and then exiled Lono. Sahlins convinces me that it was the latter. When the record says that after the killing a Hawaiian asked whether Lono would come again, Obeyesekere tries to finesse the question, while Sahlins observes how perfectly it fits his own analysis. There is an important ritual performed around Cook soon after his arrival. Obeyesekere says this is the treatment for a new chief, and that Cook was furious at being made to prostrate himself. Cook clearly had no idea what was going on, but there is no indication in any of the English texts that he was upset—and the ritual, Sahlins argues effectively, was a god-welcoming and not a chief-making one. On a return voyage, Captain Vancouver wondered why he was no longer treated as a god. He was told it is because he and his men had lunched with Hawaiian women. No god could ever do that, nor indeed any Hawaiian man: men and women do not eat together, ever.

On and on. To repeat, an outsider cannot assess the argument well except in terms of coherence and consilience, but when it comes to ethnographic data, Sahlins usually takes the day. The contest is a little closer, I thought, in explaining why sex with sailors was so in demand by the Hawaiian women. Sahlins sees it as part of the role of women in the winter ceremonies with Lono. Obeyesekere speculates that installing Cook as chief, combined with later myths, helped legitimate the relations, but he is not strong on why there should be so much more immediate interracial sex here than in other parts of the South Seas, or on the northwest coast of America (where quite a good time was had by all, but not the same good time).

Some evidence can be checked from outside, which is not always good news for Obeyesekere. For example, he quotes King (the lieutenant who carried on Cook's journal after the killing) as saying that the god "resided in us," and concludes: "It is therefore entirely possible that the installation rituals helped effect this 'residence' both in Cook and in the other gentlemen present, thereby converting them into Hawaiian chiefs." That is already strange in a Hawaiian context, where there may be no concept of gods "residing in" people, notwithstanding possession ideas elsewhere on the globe. Obeyesekere is not quoting from King's own journal, but from one of several editions published in 1784, this one edited by a canon of St. Paul's. The words are not in King's journal as published by Beaglehole. Much worse: the words do not appear in the 1784 edition cited. It does not say "reside in." It says "reside amongst," and also, "dwelled with." Mighty prepositions! Whoever wrote the text, it has to mean that the god was already amongst and with the English, in their land, not residing in anybody.

There is another feature of Obeyesekere's book which is unnerving. I mention it only because there is a lot of sloppy thinking out there in the culture wars. Obeyesekere begins his universalist stance by wondering whether any people, on being encountered by Europeans, ever took the white men for gods. Or is this a European myth, which begins, perhaps, with the Spanish arriving in Mexico, and which in a later format was swallowed by the Hawaiian natives as part of a reconstructed self-history? Excellent question. Obeyesekere reflects that he, a Sri Lankan and scholar of South-East Asian peoples, never heard any such story about his own people's meeting with the white men. To be skeptical is one thing, but to argue from the Sri Lankan experience is altogether disingenuous. Sri Lanka is historically one of the most cosmopolitan places on earth, a center of civilization when Britain was a distinctly unsceptred isle. Today, you will hardly want to sail around India without putting in at the island (and people have been sailing around India ever since sailing began). Its inhabitants were converted to Buddhism 2500 years ago; when the British finally took over, the monarchy had a genealogy as old as the one claimed for Persia. The island was known to ancient Greece. Sri Lanka was well known in Western Asia (what Europeans call the Near East) for as long as we can tell. We know that European visitors came to Sri Lanka by the ninth century at the latest. Even if those visitors were, improbably, the first visitors with pale faces, Sri Lankans must have already been pretty jaded by skin variants. Sri Lankans are hardly the ones to be surprised—let alone moved to adulation—when a new face appears on the horizon. Hawaii is a different story. I do suspect that the islanders would have known a lot more about Europeans than common wisdom and history teach. But even if my suspicions are correct, the English would have been a novelty in Hawaii, while they were no such thing in Sri Lanka.

Language and Sex

To conclude in my role as a philosopher, there is a small point where I am more universalist that Obeyesekere himself. He doubts the reports in the journals of conversations between officers and chiefs, for how could the English have understood Hawaiian? Sahlins argues convincingly that they quickly recognized morphemes similar to those of other Polynesian regions, such as Tahiti—some men aboard had a lot of linguistic experience with other South Pacific peoples who spoke related languages. Cook had with him able linguists who were especially good, it seems, on phonemes. They noticed how some consonants systematically change across parts of the island system, a fact confirmed by later linguists. Cook's instructions, in self-composed pidgin, were evidently quickly followed or resisted. Claims or hopes expressed by Hawaiians were communicated, so far as one can tell by mutual reactions. The English had real trouble with the finer details of religion and ritual, and made plain when they did not understand what was told them. But so long as there are some shared interests, two alien peoples anywhere can get to understand each other remarkably quickly on a vast range of matters that are, for both of them, practical and pragmatic. Shared interests? Cook wanted vegetables, fuel, and water: the chiefs wanted iron. On first encounter, a Hawaiian thought he could help himself: it took one sharp lesson to convey the British concept of property. Then it was trade or theft (on both sides) all the way. And a lot of doing what came naturally.

A final unserious word for universalism happens to fit Sahlins's account better than Obeyesekere's. After having had dinned into us all these years that sex is culture, let us not demand too high a cultural common factor in asking about sex between those two underprivileged groups, sailors and women. Let us apply that Humeian principle, that if you want to know what happened then, think about what happens now. Maybe the sailor-boys treated the Hawaiian girls better than the Hawaiian lads. At least the sailors would sit down to lunch with the girls, something no Hawaiian man would ever do. Does anybody remember GI's stationed in England during WWII? The English men would gladly have cooked a few of them, but the English women loved them, and married them in no small numbers.

NOTES

WORKS CITED

INDEX

NOTES

1. WHY ASK WHAT?

1. On a more technical plane than that of the social worker, the philosopher Sally Haslanger (1995, 131) writes that there is a sense in which "you and I are socially constructed." This sense is rather unenterprising—it means only that we are the way we are, to some substantial extent, because of what is attributed to us, and what we attribute to ourselves. Haslanger calls this sense of social construction "discursive."
2. Badinter (1980) is a classic study of the emergence of motherhood, with its present moral connotations, in early modern Europe. The course of motherhood-like ideas is necessarily different in different civilizations. For an analysis of "Japanese Mothers" see Ueno (1996). One might balance extreme historicizing of maternal sentiments by recalling Aristotle, who wrote that the purest example of love was the love of a mother for her infant child. Let us not throw out all cross-cultural babies when we take a historicist scrubrush to the universalist bath.
3. After a talk at the New School of Social research, Linda Nicholson (and others) urged me to emphasize the extent to which social construction has been liberating. I owe the example of motherhood to a postcard from Brydon Gombay, who wondered why I was said to be so down on social construction. I'm not.
4. www.math.tohoku.ac.jp/kuroki/Sokal/index.html
5. I have recently found the introduction to and the essays in Roger Chartier's (1997) book very helpful in this regard. Chartier, the distinguished French historian of the book and the text in European civilization, discusses relations between histories of discourse and of nondiscursive events. He is well aware that histories written at different times obey different rules of verification. The exemplars used in his book, of writing about the past, are Michel de Certeau, Norbert Elias, and Michel Foucault. That may make him sound, to some, as if he is on the side of "relativism." On the contrary, a chief problem he addresses is how the writing of history today can avoid both myth-making and forgery. Chartier expresses a sen-

sitive appreciation of two facts. First, the canons of verification for history writing change in time. Second, it is not true that anything goes.

6. I do not imply that consciousness-raising is one single activity. For an analysis in connection with gender, see Stanley and Wise (1983).
7. In preparing this section I have been greatly helped by Anandi Hattiangadi.
8. "For Beauvoir, gender is 'constructed' but implied in her formulation is an agent, a *cogito,* who somehow takes on or appropriates that gender" (Butler 1990, 9). The shudder-quotes (as I take them to be) around "constructed" are deliberate; see a remark below, about Butler's own apparent rejection of construction language.
9. Elizabeth Grosz names names (1994, 15–19). She divides relevant contributors into three groups. The first she calls egalitarians. The second she calls social constructionists, who include "probably the majority of feminist theorists today [1994]: Juliet Mitchell, Julia Kristeva, Michèle Barrett, Nancy Chodorow, Marxist feminists, psychoanalytic feminists, and all those committed to a notion of the social construction of subjectivity." "In contrast with both egalitarianism and social constructionism, a third group can be discerned. Its participants include Luce Irigaray, Hèlene Cixous, Gayatri Spivak, Jane Gallop, Moire Gatens, Vicki Kirby, Judith Butler, Naomi Schor, Monique Wittig, and many others. For them, the body is crucial to understanding woman's psychical and social existence, but the body is no longer understood as an ahistorical, biologically given, acultural object. They are concerned with the *lived body,* the body insofar as it is represented and used in specific ways in particular cultures."
10. Haslanger (1995) contrasts *constitutive* social construction with the gender-coding of individuals. The latter, she writes, is "causally socially constructed," in that social factors as well as (or opposed to) biological factors produce the coding. That is, even if I have a quite ordinary male body, the fact that I and others think of me as male has been *caused* at least in part by social factors, and indeed what counts as "ordinary" itself reflects a social norm.
11. The day that I presented parts of this chapter at the Ecole Normale Supérieure, *Le Monde*'s essay, "Les New Black Intellectuals de Harvard" obligingly wrote, "en dépit de divergences, un certain consensus s'opère à Harvard pour considérer la race comme une construction sociale." (25 April 1997, *Livres,* viii) (despite differences, there is a certain consensus at Harvard to think of race as a social construction).
12. A German-speaking audience in Zurich was perturbed at my use of the word "idea," which has troublesome implications because of its play, in translation, against the important and Kantian word *Vorstellung.* I had to emphasize that I meant to use the word "idea" in the most low-brow way imaginable, simply in order to make a contrast with what I called objects. Some philosophers would prefer the word "concept" as the generic label, but I find that its connotations are too intellectual. Moreover, English admirers of Gottlob Frege use the word in a rather technical sense derived from Frege's noun, *Begriff.* I am not making Frege's

distinction between *Begriff* and *Gegenstand,* rendered in English as Concept and Object!

13. This is not the only usage, but it is a common usage when great weight is laid on these words. J. L. Austin (1961, 85) had a fine remark: "*In vino,* possibly, '*veritas*', but in sober philosophical symposium, '*verum*'." He meant that we ought to attend to the commonplace ways in which we use the adjective, rather than the solemn ways in which we use the noun, "truth." There are also plenty of non-elevated ways in which to use the expression, "the truth"; for an example, see Chapter 8, note 2.
14. Les Jeunes et les Médias Demain, at UNESCO, Paris, 21–25 April 1997.
15. The paragraph of mine preceding the one from which I quote was quite ironic about social construction. Other mentions of social construction in the book are also deliberately distanced from the idea. Cf. pp. 116, 257. By the way, the same offending sentence which I quote above was in the original version of Chapter 5 below (Hacking 1992a, 194). In the footnote then attached to the sentence I wrote, "I do not pretend to define the word 'construction' which so many others are able to use so efficiently." I am not much better off, some seven years later.
16. For readers who have never fallen into the ambit of baseball, Fish did not mean the physical objects when he spoke of balls, but of pitches in which the ball passes outside the zone where the batter is obliged to try to hit it.

2. TOO MANY METAPHORS

1. Social construction was not, as some have suggested, a fad of the 1980s that faded fast. It continues in the 1990s, with academic computing (Shields 1995), AIDS (Clay 1995), choice (Basen et al. 1993), dementia (Harding 1997), expertise (Savage 1996), the feeble mind (Trent 1994), growth management (Innes 1991), Indian forests (Jeffrery 1998), inequality (Kelly 1993), the Landsat satellite system (Mack 1990)—no, that is not an error, a book about building satellite systems, but a book subtitled *The Social Construction of the Landsat Satellite System*—masculinity (Reynaud 1983), the nation state (Goldring 1993, and McNeely 1995), the past (Bond and Gilliam 1994), "race" (Figueroa 1991), school success (Mehan et al. 1996), and white collar crime (Savelsberg 1994). By the time one has finished collecting topics, one begins to feel a need to read *Cabbage Syndrome and the Social Construction of Dependence* (Barnes 1990).
2. I owe this brief mention of an important topic to discussion and correspondence with Kathryn Addelson, a scholar who is occupied by questions of collective responsibility on the part of professionals, including professors. See, for example, Addelson (1998).
3. This and the next four examples are taken from Carlson (1998).
4. A purely personal anecdote: Goodman came across a short paper of mine about Latour and Woolgar's *Laboratory Life* (Hacking 1988a). He immediately wrote

asking that offprints of that paper be sent to his own immediate circle of constructionalists, because Latour and Woolgar had so brilliantly presented a case of world-making. In return, I chose to call a subsequent study of a more social phenomenon (revised below as Chapter 5) "World Making by Kind-Making: Child Abuse, for example." Social studies can put historical flesh on the abstract bones of Goodman's philosophy.

3. WHAT ABOUT THE NATURAL SCIENCES?

1. "What is all the fuss about?" That is the by-line for a piece by Jean Bricmont and Alan Sokal (1997b) in *The Times Literary Supplement.* They say that the fuss is about the way in which some French writers, who became fashionable in the United States, said silly things about the sciences, often using the names of famous scientists. Bricmont and Sokal came out and said how ill-informed were the French intellectuals' remarks about science (and, they emphasized, lest anyone make wrong leaps, that Althusser, Barthes, and Foucault were not among their targets).
2. Sismondo (1993) distinguishes four distinct kinds of "social constructivism" about science, and Knorr-Cetina (1989) paints other distinctions within *Konstructivismus* on a broader canvas.
3. A good brief collective statement is Barnes and Bloor (1983). Their most recent statement of intent is Barnes, Bloor and Henry (1996). A very useful outsider's account of this program has been written by the distinguished philosopher of science, Mary Hesse (1985).
4. Barbara Herrnstein-Smith reminded me of this in e-correspondence some time ago. An anonymous referee wrote to the effect that he or she admired the distinctions in this chapter, but could not make them.
5. Arthur Fine (1996, 237) says of social constructionists that "despite occasional disclaimers, the tenor of their preaching is against science." He quotes a passage from Pickering's *Constructing Quarks* as illustration. "There is no obligation upon anyone framing a view of the world to take account of what twentieth-century science has to say . . . World views are cultural products; there is no need to be intimidated by them."

   I sympathize with Fine's doubts about Pickering's rambling passage. Contrary to Pickering, (1) there is every reason to be intimidated by quite a number of world views and cultural products on offer at present; (2) it does not make much sense to talk of someone "framing a view of the world." Grammatically that sounds like framing a view of the Matterhorn. But what does it mean? To what "obligations" must such framers submit? (3) World-view-framers who take no account of what twentieth-century science has to say sound extraordinarily arrogant, conceited, and indifferent to their fellow human beings.

   For all that, Pickering never suggested that the standard model in quantum field theory is false or dubious. Pickering probably meant that one can live a coherent and rich human life (even a holy one) without paying much attention to what twentieth-century science has to say. That is a truism. It is quite different from maintaining or even suggesting that the propositions received in the natural sciences are in general false.

6. My talk of "robust fit" is patterned after statistics. A statistical test is called robust when it leads to the same conclusion (e.g. "the data show that a hypothesis should be rejected") even when background assumptions (models) vary substantially. The idea is that a test is robust when it does not rely on the specifics of a particular model.

   The word "robust" can serve many purposes. William Wimsatt (1981) has been using it to evade philosophical debates about scientific realism. "Things [such as for example electrons] are robust if they are accessible (detectable, derivable, definable, producible, or the like) in a variety of independent ways" (Wimsatt 1994, 210f). As I understand him, he does not assert that a thing is real when it is robust. He means that we should ask not whether things, properties, relations, and larger theoretical structures are "real," but whether they are robust. Jason Robert drew my attention to this usage, parallel to, but different from, that of the statisticians.

7. For many years now American philosophers have misleadingly spoken of a "Quine-Duhem thesis," because of an analogy between Duhem's observation and Quine's view that any sentence in the fabric of belief can be revised in the light of recalcitrant experience. That way of putting things exemplifies what Quine has called semantic ascent (it talks about sentences, not sciences) and destroys Duhem's own perceptive basis for his thesis, a distinction between theories under investigation and theories about how the apparatus works. Duhem was a physicist-historian of science, while Quine is a logician-semanticist. The physicist and historian was concerned with real-life possibilities of rethinking how one's apparatus works, while the logician and semanticist was thinking in ideal terms about abstract relations between sentences. It is said that Quine was led to mention Duhem in the first place only after someone had observed a similarity to Duhem's ideas. Perhaps the central feature of Quine's doctrine boils down to holism, which might better be written "wholism"; that is, the idea that our system of beliefs must be considered as a whole. Recently Quine (1992, 14) charmingly gets the balance between himself and Duhem just right: "Pierre Duhem made much of [holism] early in this century, but not too much." I owe this quotation, and much useful criticism, to Michael Ashooh.

8. I owe this example to a conversation with Peter Galison.

9. Skuli Sigurdsson drew my attention to this passage.

10. Williams is not arguing that we know science is true because that is the best explanation of why we converge, an argument used by J. J. C. Smart and others. (And rejected by Larry Laudan and others on the ground that sciences do not tend to converge anyway.) Smart's position has been called convergent realism. The

conclusion of the argument is, "the propositions of mature sciences are probably close to the truth." That is not the conclusion of Williams's argument.

11. Some readers may find the following analogy helpful. We say that a coin is fair; perhaps that means that in an ideal limit, the relative frequency of heads will converge on ½. But the actual observed relative frequency at any point in time, after finitely many tosses, is not predetermined, even if it is increasingly probable that the frequency will be near ½. Any frequency reached at present is formally consistent with a limit of ½. Both the idea of frequencies as ideal limits, and the idea of truth as that upon which inquiry converges are due to C. S. Peirce.
12. For more on the absolute conception, see Williams 1978, 245–47; 1981, 1985, chs. 8, 9; for questions about it, see Putnam 1992, ch. 5.
13. This was one aspect of the strong program in the sociology of knowledge advanced in the 1970s by Barry Barnes and David Bloor. They wanted to understand why a body of beliefs counted as knowledge at a place and at a time for a community of knowers. Explanations of why beliefs are held should invoke, among other things, social circumstances. Scientific sociologists should not differentiate between beliefs they hold to be true, and those they hold to be false. Beliefs should be treated "symmetrically." One should not use the truth of a true belief to explain why people hold or held it—and then invoke social factors to explain why people held false or unreasonable beliefs. That would violate a principle of symmetry.

    Evidence, or reasonableness, is quite another matter from truth. Barnes and Bloor are often taken to hold a symmetry thesis about evidence: you cannot invoke the evidence available in a community for a belief in *p*, in order to explain why people in the community believed *p*. That is, we cannot invoke what *we* take to be good evidence for *p*, and their acquaintance with *p* and their taking it be good evidence for *p*, in order to explain why they believed *p*. We should not explain the fact that people held *p* by saying that it was (what we deem to be) reasonable for them to believe *p*, given the evidence that was available to them. I find this claim (about evidence, not truth) unsatisfactory.
14. This way of putting things helps answer a question I was asked in Vancouver. "Why do idealism and nominalism get so entangled?" Nominalism as explained here has affinities to Kant's transcendental idealism, except in this version there is no noumenal structure at all: it is not merely unknowable.
15. I used to attribute this saying to C. G. Jung, for I am sure I read it in some work of his. When I said so in public, in Zurich, Jung's home town, I received an amusing letter from a member of the audience: (a) suggesting that the saying came from a story by Borges, about a library in which every book attributed the saying to another book in the library, and (b), that perhaps I was Borges.
16. The sentence continues "with different laws of nature for different cultures." The contingency thesis may be confused with multiculturalism, but it has nothing to do with it. Perhaps some confused multiculturalists think that contingency has something to do with different (read oppressed) cultures or subcultures, that have

not had the privilege of doing their own thing, but that is not the point of the contingency thesis at all. The point is that "our" culture could have developed other equally successful physics. Feminists and postcolonial thinkers have indeed urged that other cultures or subcultures would do science differently, partly because their conceptions of success are different (Harding 1998). But the metaphysical contingency thesis of sticking point #1 has nothing to do with "different laws of nature for different cultures."

17. I have not found physicists claiming that constructionists have undermined their funding. Oddly, that claim was advanced a decade earlier, when the enemies were different, in an article in *Nature* written by two enraged physicists (Theocharis and Psimopoulos 1987). Prime Minister Thatcher had just curtailed funding for pure science in Britain. Who was to blame? The article began with four photographs, a portrait gallery of four rogues: Karl Popper, Thomas Kuhn, Imre Lakatos, and Paul Feyerabend. Popper?! Yes, because he said that scientific propositions are falsifiable, so the great British public no longer put its trust in science.
18. This sentence is adapted from one in a letter from Lorraine Daston, dated 17 July 1997. She made many more useful comments, which are incorporated below.
19. Oh dear. A graduate student in Toronto drew my attention to the obvious sexual connotations of this sentence, but I decided to leave it.
20. Best expressed in Sokal (1996c), originally circulated on the Internet, with the statement that it had been submitted to *Social Text,* the journal that published Sokal's original lampoon.
21. Paul Hoyningen-Huene pointed out to me after a talk at the Eidgenössische Technische Hochschule in Zurich that Kuhn was rather an inevitabilist when it came to normal science, and that he was a contingentist only for revolutions.
22. "In Feyerabend's introduction to the third edition of *Against Method,* published shortly before he died," Fuller (1995, 13) alludes to Kuhn's 1991 Rothschild Lecture at Harvard, titled "The Trouble with the Historical Philosophy of science." Weinberg also mentions this lecture. The trouble in question pertains to the Sociology of Scientific Knowledge. Feyerabend, wrote Fuller, "agreed with Kuhn that the sociologists were just as much in the wrong to criticize or demystify natural science as the natural scientists had been with respect to sociology."
23. My own reasons for my own scores are banal. I expect that I am a nominalist because I was born that way. But can I really go whole-hog with Thomas Hobbes and Nelson Goodman? No. I am only slightly tempted by contingency, because I think that the "form" our knowledge takes is contingent, but once we have asked the questions, they get answered in a fairly predetermined way, as I try to explain in Chapter 6. I rank myself as a 3 on the stability issue because my own line diverges quite strongly from both constructionists and traditionalists, although I have learned so much from both. Possibly my 1988 publication reintroduces stability into generalist philosophy of science as *the* future issue, while my 1991 work sketches an account of the stability of the laboratory sciences. In a review

of Pickering (1992) in the *Times Literary Supplement,* Lewis Wolpert dismissed my thoughts on self-vindication in the laboratory sciences as "epistemology and metaphysics," which rather cheered me, for I like those ancient trades I ply, but which meant he despised them. Most social students of knowledge, with the possible exceptions of Pickering and Latour, have an equally hard time fitting these thoughts of mine into their frameworks. Hence the ambivalent score of 3.

4. MADNESS: BIOLOGICAL OR CONSTRUCTED?

1. A research paper by David Pantalony introduced me to ADHD and its precursors, but he might not agree with my take on this series of diagnoses from fidgety to ADHD.
2. This confrontation is, not coincidentally, on the terrain of one of Latour's favorite actants. My copy of Latour's *Les Microbes* (stupidly translated as *The Pasteurization of France,* Latour 1986) was given me by Douglas; it is the very copy that Latour had given to her.
3. Andrew Pickering (forthcoming) uses similar observations to more dramatic effect. He thinks, contra Latour (1993), that we have been thoroughly modern. He argues that the free-standing machine is the mark of the modern, and that we begin a new era—now—when we are forced to realize that machines, people, and the old nature are kin, sharing an interlinked ecology.
4. "For the exogenously extended organizational complex functioning as an integrated system *unconsciously,* we propose the name 'Cyborg.' " (Clynes and Kline 1960/1996a, 31, emphasis added.)
5. I owe this discussion to a question posed in Paris by Pierre-Henri Castel, and clarified for me by Daniel Andler.
6. Here I owe a great deal to Licia Carlson (1998), but she may dissent from my reading of the feeble mind as a series of interactive kinds.
7. Jack and Jill are shown a box with plastic dinosaurs in it. Jack is sent out of the room. The dinosaurs are replaced by candies. Jack is asked to come back into the room, but before he enters Jill is asked, what will Jack think is in the box? If Jill says dinosaurs, she has a theory of mind, but if she says candies, she does not.
8. Spitzer has been the editor in chief of the *DSM (The Diagnostic and Statistical Manual of Mental Disorders)—DSM-III* (1980), *DSM-III(R)* (1987), and *DSM-IV* (1994). I take the anecdote from a talk by Spitzer to the annual convention of the American Psychological Association, Toronto, 10 August 1996.
9. That, at any rate, is the "official" story, recounted by Ellenberger (1970). I have been told by Swiss psychiatrists that Bleuler's career is not quite as anodyne as this Swiss psychiatrist and historian of psychiatry makes it out to be.
10. The example occurs in a number of papers of Putnam's, starting no later than his 1961 talk to the American Association for the Advancement of Science, printed in Butler (1965).

11. For example, my experimental or entity realism about the theoretical natural sciences makes heavy use of Putnam's idea; see my 1983 publication.

5. KIND-MAKING: THE CASE OF CHILD ABUSE

1. The complete court records were translated by Georges Bataille (1959), a celebrated scholar, artist, and, one might say, artistic pornographer, whose fascination with the case cannot have been in the least innocent. The ecclesiastical court excommunicated Gilles for "heretical apostasy . . . evocation of demons . . . and vice against nature with children of the one or the other sex according to the sodomite practice," and the civil court found him guilty of 400 child murders. Sound like a good case of witchcraft sex abuse? Not quite, for the evocation of demons was in connection with alchemy, through which Gilles attempted to recoup his squandered fortunes. "I did my deeds for blood's sake, not the Devils . . . lust, not necromancy." The accusation of satanism in connection with the children was rejected.
2. I take the riddle more seriously than most readers—for example, in my work of 1993a, 1993b, 1994.
3. Dorothy Smith, of the Ontario Institute for Studies of Education, first suggested that I take a look at child abuse as an example. I did not know what I was going to get into, although she, wise and puckish both, undoubtedly did. My thanks are mixed with curses.
4. *New York Times,* June 28, 1990. National edition page A13.
5. See "Making and molding" (Hacking 1991a), 264–266, for the argument.
6. For a discussion of the connection between child abuse and pollution (including self-abuse, namely masturbation), see "Making and molding" (Hacking 1991a), 277–280.
7. See *Rewriting the Soul* (Hacking 1995b, 64–66) for a statistical update.
8. One aspect of the backlash is the topic of Chapter 8 of *Rewriting the Soul* (Hacking 1995b).
9. In the 1970s the National Center for Child Abuse and Neglect was annually citing about 2,000 deaths; see Gelles 1979, 11. In 1989 it was considered that at least 1,200 and possibly as many as 5,000 American children died from abuse and neglect, *New York Times,* June 28, 1990, A13. Compared to the figures about to be cited, this is relatively constant.
10. National Center for Child Abuse and Neglect; cf. *New York Times,* April 17, 1983, A1, and *ibid.,* June 28, 1990, A13.
11. For more information on the reception of child abuse in Europe, see the longer version of this chapter (Hacking 1992a, 210–213).
12. "Speaking out," an interview with founders of Incest Crisis Line, *Sunday Times Magazine,* 9 August 1987, 10.
13. Seen on the French network FR3, October 20, 1990. The statement was in English.

The network, perhaps maliciously, added that the government in exile had just acquired an American public relations adviser.

14. " 'Tree abuse!' planter barks," *Globe & Mail*, Toronto, 22 October 1987.
15. For more discussion of child pornography as child abuse, see the longer version of this chapter (Hacking 1992a, 219–223). Internet pornography has, however, made that account appear very dated.
16. Joan Barfoot reviewing *Caesars of the Wilderness* by Peter Newman, *New York Times Book Review*, 20 December 1987, p. 9.
17. Chapter 16 of *Rewriting the Soul* (Hacking 1995b) discusses the consequences of this idea in depth.

6. WEAPONS RESEARCH

1. Estimates of Research and Development spending are notoriously inaccurate. A possibly extreme version, current at the time this essay was first published, was in the *Bulletin of the Atomic Scientists* 42, 3 (1986). Using *National Science Foundation Report* 85-322, the writer inferred that military spending accounted for 72.7 percent of total American public R&D investment. To get a sense of the relevant sums, observe that in 1986 the Strategic Defense Initiative ("star wars") alone had about the same budget as the National Institutes of Health, namely slightly less than $5 billion. But figures have to be interpreted. A good deal of work that was already funded was transferred to SDI accounts. Although this was only a paper transaction, it ensured direct responsibility to the military for some work that had not been primarily military in nature.

7. ROCKS

1. I first learned about sedimentology and nanobacteria at a 1997 lecture by Judith McKenzie at the Swiss Federal Polytechnic (Eidgenössische Technische Hochschule) in Zurich. Subsequent discussion with her helped fill in the details.
2. Those who disliked my calling "truth" an elevator word (in Chapters 1 and 3) may make fun of this pair of sentences. They are deliberate. They exemplify what is misleadingly called a redundancy use of the word "truth." My paired sentences are of this form: (1) Claim *X* is closer to the truth than claim *Y*. The truth is *Z*. The force of (1) is: (2) *Z; X* is closer to *Z* than *Y*. The expression "the truth" is thus redundant. Stylistically, (2) is clumsy, while I hope that (1) reads fairly well, despite the deliberate iteration of "the truth." In this redundancy usage, "the truth" is no more than a shorthand way of referring to a statement that is also asserted, namely *Z*.

   I have no use for those who put words like "true" and "truth" in ironical shudder quotes to indicate that the speaker has been liberated from such a discredited idea as truth. That is just as bad as treating the truth with unquestioned reverence, and shows just as little respect for our shared means of communication, the English language. If I assert that dolomite is being formed in small quantities in disagreeable places such as a salt lagoon off the coast of Brazil, I have no qualms in saying that that is true. The trouble comes when the expression "the truth" becomes elevated. Maxim: if, in a philosophical discussion, you become tempted to engage in semantic ascent in order to make some point you think is important, stop, and try doing the thinking at ground level.

3. I was (out of mere prejudice) completely skeptical about that rock from the day it was announced, and I even suggested, only half in jest, that we might have another Piltdown man on our hands.
4. I failed Geology 200 at the University of British Columbia because I could not be bothered to learn how to recognize all those rocks; instead, in order to work my way through college, I became a geophysicist who interprets wiggly lines on a seismic readout. I became a whizz at a now obsolete technology, "picking" as we called it, the Mississippian and the Devonian limestone and dolomite in Alberta, roughly a mile and two miles below the earth's surface. That was a theory-laden bunch of observations if ever there was one.
5. By coincidence the Ecole des Mines is Latour's own home base, but it is not well networked in France with respect to philosophical thinking about the sciences. Latour does have a small and significant network in France, but his larger network is primarily located in the English-speaking world, to the extent that *Science in Action* was first published in English.
6. McKenzie had never heard of them when I mentioned them at lunch in May 1997.

8. THE END OF CAPTAIN COOK

1. Retroactive diagnosis, as I have repeatedly said elsewhere, is a mug's game, so this remark is only half-serious. We think of Alzheimer's in terms of memory loss. It is very striking to read in the psychiatric records from early in the century, when the neurologist Alois Alzheimer identified the plaque that is associated with the disorder, that memory loss is not the primary sign of dysfunction. This is especially true when the illness strikes early, that is, when the patients are in their forties. Instead we encounter meaningless irritability and aggression, combined with confusion. One of the reasons that loss of memory is so emphasized today is that we can easily define objective quantitative tests for memory loss, but have no agreed way to measure degree of aggression. This observation could well have been a footnote for Chapter 4! Early-onset Alzheimer's seems to run in families. Cook's ancestors show no signs of it, so far as we can tell. Maybe he just cracked up.

WORKS CITED

AAMR. 1992. *Handbook of the American Association for the Mentally Retarded,* X, 9.

Addelson, Kathryn Pine. 1998. The birth of the fetus. In *The Fetal Imperative: Feminist Positions,* ed. Meredith Michaels and Lynn Morgan. Philadelphia: University of Pennsylvania Press.

Allen, Jeffner. 1989. Women who beget women must thwart major sophisms. In *Women, Knowledge, and Reality,* ed. Ann Garry and Marilyn Pearsall. Boston: Unwin Hyman.

Anastasi, A. 1988. *Psychological Testing.* 6th ed. New York: Macmillan.

Andrews, George Reid. 1995. *The Social Construction of Democracy, 1870–1990.* New York: New York University Press.

Anson, R. 1980. The last porno show. In *The Sexual Victimology of Youth,* ed. L. G. Schultz. Springfield, Ill.: Charles C. Thomas.

Appiah, Kwame Anthony, and Amy Gutmann. 1996. *Color Conscious: The Political Morality of Race.* Princeton: Princeton University Press.

Ariès, Philippe. 1962. *Centuries of Childhood: A Social History of Family Life,* trans. Robert Baldick. New York: Knopf.

Arney, William Ray, and Bernard J. Bergen. 1984. Power and visibility: The invention of teenage pregnancy. *Social Sciences and Medicine* 18: 11–19.

Asche, Adrienne and Michelle Fine. 1988. Introduction: Beyond pedestals. In *Women with Disabilities: Essays in Psychology, Culture, and Politics,* ed. Adrienne Asche and Michelle Fine. Philadelphia: Temple University Press.

Austin, John Langshaw. 1961. *Philosophical Papers.* Oxford: Clarendon Press.

——— 1962. *Sense and Sensibilia.* Oxford: Clarendon Press.

Averill, J. R. 1980. A constructivist view of emotions. In *Theories of Emotion,* ed. R. Plutchik and H. Kellerman. New York: Academic Press.

Badgely, Robin M. 1984. *Sexual Offences against Children.* Report of the Committee on Sexual Offences against Children and Youths Appointed by the Minister of Justice and Attorney-General of Canada and the Minister of Health and Welfare. Ottawa: Supply and Services Canada.

Barnes, Barry. 1977. *Interests and the Growth of Knowledge.* London: Routledge and Kegan Paul.

——— 1995. *The Elements of Social Theory.* London: UCL Press.

Barnes, Barry, and David Bloor. 1982. Relativism, rationalism and the sociology of knowledge. In *Rationality and Relativism,* ed. Martin Hollis and Steven Lukes. Oxford: Blackwell.

Barnes, Barry, David Bloor, and John Henry. 1996. *Scientific Knowledge: A Sociological Analysis.* Chicago: University of Chicago Press.

Barnes, Colin Greenhill. 1990. *Cabbage Syndrome and the Social Construction of Dependence.* London: Falmer.

Barrett, Edward, ed. 1992. *Sociomedia: Multimedia, Hypermedia, and the Social Construction of Knowledge.* Cambridge, Mass.: MIT Press.

Basen, Gwynne et al. 1993. *Misconceptions: The Social Construction of Choice and the New Reproductive and Genetic Technologies.* Hull, Québec: Voyageur.

Bataille, Georges. 1959. *Le procès de Gilles de Rais.* Paris: J.-J. Pauvert.

Beaglehole, John Cawte, ed. 1968. *The Journals of Captain James Cook on His Voyages of Discovery.* Cambridge: Cambridge University Press for the Hakluyt Society.

Beard, Mary. 1990. Review of Tate (1990). *Times Literary Supplement,* 14 September: 968.

Beck, Lewis White. 1950. Constructions and inferred entities. *Philosophy of Science* 17: 74–86.

Bell, Stuart. 1988. *When Salem Came to the Boro: The True Story of the Cleveland Child Abuse Case.* London: Bell.

Berger, Peter, L., and Thomas Luckmann. 1966. *The Social Construction of Reality: A Treatise in the Sociology of Knowledge.* Garden City, N.Y.: Doubleday.

Bergman, Abraham B. 1986. *The "Discovery" of Sudden Infant Death Syndrome: Lessons in the Practice of Political Medicine.* New York: Praeger.

Besharov, D. J. 1985. "Doing something" about child abuse: The need to narrow the grounds for state intervention. *Harvard Journal of Law and Public Policy* 8: 538–589.

Bhattacharya, A. K. 1979. Child abuse in India and the nutritionally battered child. *Child Abuse and Neglect: The International Journal* 3: 607–614.

Bijker, Wiebe E., Thomas P. Hughes, and Trevor Pinch, eds. 1987. *The Social Construction of Technological Systems: New Directions in the Sociology and History of Technology.* Cambridge, Mass.: MIT Press.

Bishop, E. 1967. *Constructive Analysis.* New York: McGraw-Hill.

Bloor, David. 1976. *Knowledge and Social Imagery.* London: Routledge.

Bond, George Clement, and Angela Gilliam. 1994. *The Social Construction of the Past: Representation as Power.* London: Routledge.

Boyle, Mary. 1990. *Schizophrenia: A Scientific Delusion?* London: Routledge.

Bricmont, Jean, and Alan Sokal. 1997a. *Impostures intellectuelles.* Paris: Odile Jacob.

——— 1997b. What is all the fuss about? *Times Literary Supplement,* 17 October: 17.

Browning Dianne H., and Bonny Boatman. 1977. Incest: Children at risk. *American Journal of Psychiatry* 134: 69–72.

Butler, Judith. 1990. *Gender Trouble: Feminism and the Subversion of Identity.* New York: Routledge.

Butler-Sloss, Elizabeth. 1988. *Report of the Inquiry into Child Abuse in Cleveland 1987.* London: Her Majesty's Stationery Office.

Byung, Hoon Chun. 1989. Child abuse in Korea. *Child Welfare* 68: 154–159.

Campbell, Beatrix. 1987. The skeleton in the family's cupboard. *New Statesman,* 31 July: 11.

——— 1988. *Unofficial Secrets: Child Sexual Abuse—The Cleveland Case.* London: Virago.

Capps, Lisa, and Elinor Ochs. 1995. *Constructing Panic: The Discourse on Agoraphobia.* Cambridge, Mass.: Harvard University Press.

Carlson, Licia. 1998. Mindful subjects: Classification and cognitive disability. Ph.D. diss., University of Toronto.

Carnap, Rudolf. 1928/1967. *The Logical Structure of the World,* trans. Rolf A. George. Berkeley: University of California Press.

Chapman, William. 1991. *Inventing Japan: The Making of a Postwar Civilization.* New York: Prentice Hall.

Chartier, Roger. 1997. *On The Edge of the Cliff: History, Language, and Practices.* Baltimore: Johns Hopkins University Press.

Clawson, Mary Ann. 1989. *Constructing Brotherhood: Class, Gender, and Fraternalism.* Princeton: Princeton University Press.

Clay, C. J. 1995. *The Social Construction of AIDS.* Sheffield: Eng.: Sheffield City Polytechnic.

Clynes, Manfred E., and Nathan S. Cline. 1960/1995. Cyborgs and space. *The Cyborg Handbook,* eds. Chris Hables Gray. New York: Routledge.

Code, Lorraine. 1995. *Rhetorical Space: Essays on Gendered Allocations.* New York: Routledge.

Cohen, Robert S., and Thomas Schnelle. 1986. *Cognition and Fact: Materials on Ludwik Fleck.* Dordrecht: Reidel.

Collins, Harry M. 1985. *Changing Order: Replication and Induction in Scientific Practice.* London: Sage.

——— 1990. *Artificial Experts: Social Knowledge and Intelligent Machines.* Cambridge, Mass.: MIT Press.

——— 1993. On Lewis Wolpert's *The Unnatural Nature of Science.* Public Understanding of Science 2: 261–264.

——— 1995. Being and becoming. *Nature* 376, 13 July: 131. A review of Searle (1995).

——— 1998. The meaning of data: Open and closed evidential cultures in the search for gravitational waves. *American Journal of Sociology* 104: 293–337.

Collins, Harry M., and Trevor J. Pinch. 1982. *Frames of Meaning: The Social Construction of Extraordinary Science.* London: Routledge & Kegan Paul.

——— 1993. *The Golem: What Everyone Should Know about Science.* Cambridge: Cambridge University Press.

Comaroff, Jean. 1994. Aristotle re-membered. In *Questions of Evidence: Proof, Practice, and Persuasion across the Disciplines,* ed. James Chandler, Arnold I. Davidson, and Harry Harootunian. Chicago: University of Chicago Press, 1994.

Connor, Steve. 1996. Small wonder. *Sunday Times,* 21 July, 14.

Cook-Gumperz, Jenny, ed. 1986. *The Social Construction of Literacy.* Cambridge: Cambridge University Press.

Coulter, Jeff. 1979. *The Social Construction of Mind: Studies in Ethnomethodology and Linguistic Philosophy.* Totawa, N.J.: Rowman and Littlefield.

Danziger, Kurt. 1990. *Constructing the Subject: Historical Origins of Psychological Research.* Cambridge: Cambridge University Press.

Daston, Lorraine. 1992. Objectivity and the escape from perspective. *Social Studies of Science* 22: 567–618.

Daston, Lorraine, and Peter Galison. 1992. The image of objectivity. *Representations* 40: 81–128.

Davenport, W. H. 1976. Sex in cross-cultural perspective. In *Human Sexuality in Four Perspectives,* ed. Frank A. Beach. Baltimore: Johns Hopkins University Press.

Davy, Humphry. 1805/1980. *Humphry Davy on Geology: The 1805 Lectures for a General Audience.* Edited with an introduction by Robert H. Siefried and Robert H. Dott Jr. Madison: University of Wisconsin Press.

——— 1812. *Elements of Chemical Philosophy.* London: J. Johnson.

de Beauvoir, Simone. 1953. *The Second Sex,* trans. H. M. Parshley. New York: Knopf.

deMause, Lloyd. 1974. The evolution of childhood. In deMause, *The History of Childhood: The Untold Story of Child Abuse.* New York: Psychohistory Press. 1–73.

——— 1988. The universality of incest. *Journal of Psychohistory* 15.

Dewar, Alison MacKenzie. 1986. The social construction of gender in a physical education programme. Master's thesis, University of British Columbia.

De Young, Mary. 1988. The good touch/bad touch dilemma. *Child Welfare* 67: 60–68.

Donellan, Anne M., and Martha A. Leary. 1995. *Movement Differences and Diversity in Autism/Mental Retardation: Appreciating and Accommodating People with Communication and Behavior Challenges.* Madison, Wis.: DRI Press.

Donovan, Denis M. 1991. Darkness invisible. *Journal of Psychohistory* 19: 165–184.

Donzelot, Jacques. 1979. *The Policing of Families,* trans. Robert Hurley. New York: Macmillan.

Douglas, Jack D. 1970. *Deviance and Respectability: The Social Construction of Moral Meanings.* New York: Basic Books.

Douglas, Mary. 1986. *How Institutions Think.* Syracuse: Syracuse University Press.

Douglas, Mary, and Aaron Wildavsky. 1982. *Risk and Culture: An Essay on the Selection of Technical and Environmental Dangers.* Berkeley: University of California Press.

Duhem, Pierre. 1906/1954. *The Aim and Structure of a Physical Theory.* Princeton: Princeton University Press.

Dyson, George. 1997. *Darwin among the Machines.* Reading, Mass.: Addison Wesley.

Eder, Klaus. 1996. *The Social Construction of Nature: A Sociology of Ecological Enlightenment.* Introduction and parts 1 and 2 translated by Mark Ritter. London: Sage.

Eekelaar, John M., and Sanford N. Katz, eds. 1978. *Family Violence: An International and Interdisciplinary Study.* Toronto: Butterworths.

Ekman, Paul. Afterword to Charles Darwin, *The Expression of the Emotions in Man and Animals,* ed. Ekman. Oxford: Oxford University Press.

Ellenberger, Henri. 1970. *The Discovery of the Unconscious: The History and Evolution of Dynamic Psychiatry.* New York: Basic Books.

Emery, George Neil. 1993. *Facts of Life: The Social Construction of Vital Statistics, Ontario, 1859–1952.* Montreal: McGill–Queens University Press.

Ernest, Paul. 1998. *Social Constructivism as a Philosophy of Mathematics.* Albany: State University of New York Press.

Ferrier, Pierre. 1986. Presidential address to The International Society for the Prevention of Child Abuse and Neglect. *Child Abuse and Neglect. The International Journal* 10: 279.

Feyerabend, Paul. 1987. Notes on relativism. In Feyerabend, *Farewell to Reason.* London: Verso.

——— 1994. Potentially every culture is all cultures. *Common Knowledge* 3, no. 2: 16–22.

Feynman, Richard P. 1967. *The Character of Physical Law.* Cambridge, Mass.: MIT Press.

Figueroa, Peter M. E. 1991. *Education and the Social Construction of "Race."* London: Routledge.

Fine, Arthur. 1986. *The Shaky Game: Einstein, Realism, and the Quantum Theory.* Chicago: University of Chicago Press.

——— 1996. Science made up. In *The Disunity of Science: Boundaries, Contexts, and Power,* ed. Peter Galison and David Stump. Stanford: Stanford University Press.

Finkelhor, David. 1979a. *Sexually Victimized Children.* New York: Free Press.
——— 1979b. What's wrong with sex between adults and children? Ethics and the problem of sexual abuse. *American Journal of Orthopsychiatry* 49: 692–697.
——— 1986. *A Source Book on Child Sexual Abuse,* Beverly Hills: Sage.
Fish, Stanley. 1996. Professor Sokal's bad joke. *New York Times,* 21 May, op-ed page.
Fleck, L. 1935/1979. *Genesis and Development of a Scientific Fact.* Chicago: University of Chicago Press.
Folk, R. L. 1993. SEM imaging of bacteria and nannobacteria in carbonate sediments and rocks. *Journal of Sedimentary Petrology* 63: 990–999.
Ford, Clellan S., and Ford A. Beach. 1952. *Patterns of Sexual Behavior.* London: Eyre & Spottiswoode.
Forman, Paul. 1987. Beyond quantum electronics: National security as a basis for physical research in the United States, 1940–1960. *Historical Studies in the Physical and Biological Sciences* 18: 149–229.
Forward, Susan, and Craig Buck. 1979. *Betrayal of Innocence: Incest and Its Devastation.* Harmondsworth: Penguin.
Fowler, H. W. 1926. *A Dictionary of Modern English Usage.* Oxford: Clarendon Press.
Fowlkes, Martha R. 1982. *Love Canal: The Social Construction of Disaster.* Washington, D.C.: Federal Emergency Management Agency.
Fraser, Gertrude, and Philip L. Kilbride. 1980. Child abuse and neglect—rare but perhaps increasing phenomena among the Samia of Kenya. *Child Abuse and Neglect: The International Journal* 4: 227–232.
Fraser, Nancy. 1989. *Unruly Practices: Power, Discourse, and Gender in Contemporary Social Theory.* Minneapolis: University of Minnesota Press.
Fuller, Steve. 1995. Paul Feyerabend (1924–1994): An appreciation. *Vest* 8: 7–15.
Galison, Peter. 1987. *How Experiments End.* Chicago: University of Chicago Press.
——— 1990. Aufbau/Bauhaus: Logical positivism and architectural modernism. *Critical Inquiry* 16: 709–752.
——— 1995. Context and constraint. In *Scientific Practice: Theories and Stories of Doing Physics,* ed. Jed. Z. Buchwald. Chicago: University of Chicago Press.
——— 1997. *Image and Logic: A Material Culture of Microphysics.* Chicago: University of Chicago Press.
Garcia, Claudia. 1996. *Making of the Miskitu People of Nicaragua: The Social Construction of Ethnic Identity.* Uppsala: Almqvist and Wiksell.
Gell-Mann, Murray. 1994. *The Quark and the Jaguar: Adventures in the Simple and the Complex.* New York: W. H. Freeman.
Gelles, Richard J. 1975. The social construction of child abuse. *American Journal of Orthopsychiatry* 45: 363–371.
——— 1979. *Family Violence.* Beverly Hills: Sage.
Gelles, Richard J., and Claire Patrick Cornell. 1983. Introduction: An international

perspective on family violence. In Gelles and Cornell, *International Perspectives on Family Violence.* Lexington, Mass.: Lexington Books.

Gil, David C. 1968. Incidence of child abuse and neglect: demographic characteristics of persons involved. In Helfer and Kempe (1968).

——— 1975. Unravelling child abuse. *American Journal of Orthopsychiatry* 45: 346–356.

——— 1978. Societal violence in families. In Eekelaar and Katz (1978).

Gingras, Yves, and S. S. Schweber. 1986. Constraints on construction. *Social Studies of Science* 16: 372–383.

Giovannoni, J. M., and R. M. Becerra. 1979. *Defining Child Abuse.* New York: Free Press.

Glashow, Sheldon. 1992. The Death of Science!? In *The End of Science? Attack and Defense,* ed. Richard J. Elvee. Lanham, Md.: University Press of America.

Goddard, Henry Herbert. 1927. *Two Souls in One Body? A Case of Dual Personality. A Study of a Remarkable Case: Its Significance for Education and for the Mental Hygiene of Childhood.* New York: Dodd Mead.

Goffman, Erving. 1963. *Stigma: Notes on the Management of Spoiled Identity,* Englewood Cliffs, N.J.: Prentice-Hall.

Golan, Dafnah. 1994. *Inventing Shaka: Using History in the Construction of Zulu Nationalism.* Boulder: L. Rienner.

Goldring, Maurice. 1993. *Pleasant the Scholar's Life: Irish Intellectuals and the Construction of the Nation State.* London: Serif.

Goode, Erich. 1994. *Moral Panics: The Social Construction of Deviance.* Cambridge, Mass.: Blackwell.

Gooding, David. 1990. *Experiment and the Making of Meaning: Human Agency in Scientific Observation and Experiment.* Dordrecht: Kluwer.

Goodman, Nelson. 1940/1990. *A Study of Qualities.* New York: Garland.

——— 1951. *The Structure of Appearance.* Cambridge, Mass.: Harvard University Press.

——— 1954/1983. *Fact, Fiction, and Forecast.* Indianapolis: Hackett.

——— 1978. *Ways of Worldmaking.* Indianapolis: Hackett.

Goodman, Nelson, and Willard van Orman Quine. 1947. Steps towards a constructive nominalism. *Journal of Symbolic Logic* 12: 97–122.

Gould, Stephen Jay. 1977. *Ever Since Darwin: Reflections in Natural History.* New York: W. W. Norton.

Grandin, Temple. N.d. My experiences with visual thinking, sensory problems, and communication difficulties. Typescript, Colorado State University, Fort Collins.

Greenland, Cyril. 1988. *Preventing C.A.N. Deaths: An International Study of Deaths Due to Child Abuse and Neglect.* London: Routledge, Chapman & Hall.

Griffiths, Paul E. 1997. *What Emotions Really Are: The Problem of Psychological Categories.* Chicago: University of Chicago Press.

Grosz, Elizabeth. 1994. *Volatile Bodies: Toward a Corporeal Feminism.* Bloomington: Indiana University Press.

Grünzweig, Walter, and Roberta Maeirhofer, eds. 1992. *Constructing the Eighties: Versions of an American Decade.* Tübingen: Gunter Narr Verlag.

Hacking, Ian. 1975. *Why Does Language Matter to Philosophy?* Cambridge: Cambridge University Press.

——— 1982. Language, truth and reason. In *Rationality and Relativism,* ed. M. Hollis and S. Lukes. Oxford: Blackwell.

——— 1983. *Representing and Intervening.* Cambridge: Cambridge University Press.

——— 1986. Making up people. In *Reconstructing Individualism,* ed. Thomas Heller, Morton Sosna, and David E. Wellberry. Stanford: Stanford University Press.

——— 1988a. The participant irrealist at large in the laboratory. *British Journal for the Philosophy of Science* 39: 277–294.

——— 1988b. On the stability of the laboratory sciences. *Journal of Philosophy* 85: 507–514.

——— 1991a. The making and molding of child abuse. *Critical Inquiry* 17: 253–288.

——— 1991b. The self-vindication of the laboratory sciences. In Pickering (1991).

——— 1992a. World-making by kind-making: Child abuse for example. In *How Classification Works: Nelson Goodman among the Social Sciences,* ed. Mary Douglas and David Hull. Edinburgh: Edinburgh University Press, 180–238.

——— 1992b. "Style" for historians and philosophers. *Studies in History and Philosophy* 23: 1–20.

——— 1993a. On Kripke's and Goodman's uses of "grue." *Philosophy* 68: 269–295.

——— 1993b. Goodman's new riddle is pre-Humian. *Revue internationale de philosophie* 46: 229–243.

——— 1994. Entrenchment. In *GRUE: The New Riddle of Induction,* ed. David Stalker. La Salle, Ill.: Open Court.

——— 1995a. The looping effects of human kinds. In *Causal Cognition: A Multidisciplinary Approach,* ed. Dan Sperber, David Premack, and Ann J. Premack. Oxford: Clarendon Press.

——— 1995b. *Rewriting the Soul: Multiple Personality and the Sciences of Memory.* Princeton: Princeton University Press.

——— 1997. John Searle's building blocks. *History of the Human Sciences.*

——— 1998a. *Mad Travelers: Reflections on the Reality of Transient Mental Illnesses.* Charlottesville: Va.: University Press of Virginia.

——— 1998b. Canguilhem amid the cyborgs. *Economics and Society* 27: 202–216.

——— 1999. Teenage pregnancy: social construction? In *Early Parenting as a Social and Ethical Issue,* ed. David Checkland and James Wong. Toronto: University of Toronto Press.

Hanly, Charles. 1987. Review of Masson. 1984. *International Journal of Psychoanalysis* 67: 517–521.

Hanson, Norwood Russell. 1961. Are wave mechanics and matrix mechanics equivalent theories? In *Current Issues in the Philosophy of Science,* ed. Herbert Feigl and Grover Maxwell. New York: Holt Rinehart and Winston.

Haraway, Donna. 1985/1991. A cyborg manifesto: Science, technology and socialist-feminism in the late twentieth century. In Haraway (1991).

——— 1989. *Primate Visions: Gender, Race and Nature in the World of Modern Science.* London: Verso.

——— 1991. *Simians, Cyborgs and Women: The Reinvention of Nature.* New York: Routledge.

——— 1997. *Modest Witness*

Second Millenium.FemaleMan© Meets OncoMouseTM: Feminism and Technoscience. New York: Routledge.

Harding, Nancy. *The Social Construction of Dementia: Confused Professionals?* London: Kingsley Publishers.

Harré, Rom. 1986. *The Social Construction of Emotions.* Oxford: Blackwell.

Hartley, Gillian M., and Susan Gregory, eds. 1991. *Constructing Deafness.* London: Pinter Publications.

Haslanger, Sally. 1995. Ontology and social construction. *Philosophical Topics* 23: 127–157.

Helfer, Ray E. 1968. The responsibility and role of the physician. In Helfer and Kempe (1968).

Helfer, Ray E., and C. Henry Kempe, eds. 1968. *The Battered Child.* Chicago: University of Chicago Press.

Hempel, C. G. 1966. *Philosophy of Natural Science.* Englewood Cliffs, N. J.: Prentice Hall.

Herman, Judith, and L. Hirschman. 1977. Father-daughter incest. *Signs* 2: 735–756.

Herrnstein, Richard J., and Charles Murray. 1994. *The Bell Curve: Intelligence and Class Structure in American Life.* New York: The Free Press.

Hesse, Mary. 1985. *Revolutions and Reconstruction in the Philosophy of Science.* Brighton and Hove, Eng.: Harvester.

Hirschfeld, Lawrence A. 1996. *Race in the Making: Cognition, Culture and the Child's Construction of Human Kinds.* Cambridge, Mass.: MIT Press.

Hitchens, Christopher. 1995. *The Missionary Position: The Mother Teresa in Theory and Practice.* London: Verso.

Hobbs C. J., and J. M. Wynne. 1986. Buggery in childhood—a common syndrome of child abuse. *The Lancet* 4, October: 792–795.

Hutson, Susan, and Mark Liddiard. 1994. *Youth Homelessness: The Construction of a Social Issue.* Houndmills, Eng.: Macmillan.

IMGAC (International Molecular Genetic Study of Autism Consortium). 1998. A full genome screen for linkage to a region on chromosome 7q. *Human Molecular Genetics* 7: 571–578.

Innes, Judith. 1991. *Group Processes and the Social Construction of Growth Management: The Cases of Florida, Vermont and New Jersey.* Berkeley: University of California Institute of Urban and Regional Development.

Janko, Susan. 1994. *Vulnerable Children, Vulnerable Families: The Social Construction of Child Abuse.* New York: Teachers College Press.

Jardine, Nicholas. 1991. *The Scenes of Inquiry: On the Reality of Questions in the Sciences.* Oxford: Clarendon Press.

Jeffery, Roger. 1998. *The Social Construction of Indian Forests.* Edinburgh: Centre for East Asian Studies.

Jenkins, P. 1994. *Using Murder: The Social Construction of Serial Homicide.* New York: A. de Gruyter.

Jenneret, Yves. 1998. *L'affaire Sokal ou la querelle des impostures.* Paris: Presses Universitaires de France.

Johnson, Michael P., and Karl Hufbauer. 1982. Sudden infant death syndrome as a medical research problem since 1945. *Social Problems* 30: 65–81.

Jurdant, Baudoin (ed.). 1998. *Impostures scientifiques: Les malentendus de l'affaire Sokal.* Paris: La Découverte.

Kajander, Olavi, and Çiftoçioglu. 1998. Nanobacteria: an alternative mechanism for pathogenic intra- and extracellular calcification and stone formation. *Proceedings of the National Academy of Sciences* 95: 8274–5479.

Kamerman, Sheila B. 1975. Eight countries: Cross national perspectives on child abuse and neglect. *Children Today* 4: 36.

Keller, Evelyn Fox. 1992. Critical silences in scientific discourse: Problems of form and re-form. In *Secrets of Life, Secrets of Death: Essays on Language, Gender and Science,* ed. Evelyn Fox Keller. New York: Routledge.

Kelly, Raymond C. 1993. *Constructing Inequality: The Fabrication of a Hierarchy Virtue Among the Etero.* Ann Arbor: University of Michican Press.

Kempe, C. Henry, et al. 1962. The battered child syndrome. *Journal of the American Medical Association* 181/1.

Kempe C. Henry, et al., eds. 1980. *The Abused Child in the Family and in the Community: Selected Papers from the Second International Conference on Child Abuse and Neglect.* Oxford: Pergamon.

Kinsey, Alfred C., et al. 1953. *Sexual Behavior in the Human Female.* Philadelphia: W. B. Saunders.

Kinsman, Gary William. 1983. *The Social Construction of Homosexual Culture.* Master's thesis, University of Toronto.

Kirch, Patrick Vinton and Marshall Sahlins. 1992. *Anahulu: The Anthropology of History in the Kingdom of Hawaii.* 2 vols. Chicago: University of Chicago Press.

Kitzinger, Celia. 1987. *The Social Construction of Lesbianism.* London: Sage.

Knorr-Cetina, Karin. 1989. Spielarten des Konstructivismus. *Soziale Welt* 40: 86–96.

Knorr-Cetina, Karin, and Michael Mulkay, eds. 1983. *Science Observed: Perspectives on the Social Study of Science.* London: Sage.

Koertge, Noretta (ed.). 1998. *A House Built on Sand: Exposing Postmodernist Myths about Science.* New York: Oxford University Press.

Kripke, Saul. 1980. *Naming and Necessity.* Cambridge, Mass.: Harvard University Press.

Kuhn, Thomas. 1977. *The Essential Tension,* Chicago: University of Chicago Press.

La Fontaine, Jean. 1998. *Speak of the Devil: Allegations of Satanic Abuse in Britain.* Cambridge: Cambridge University Press.

Lakatos, Imre. 1970. Falsification and the methodology of scientific research programmes. In *Criticism and the Growth of Knowledge,* ed. Imre Lakatos and Alan Musgrave. Cambridge: Cambridge University Press.

Lakoff, George. 1986. *Women, Fire and Dangerous Things: What Categories Reveal about the Mind.* Chicago: University of Chicago Press.

Landis, Judson T. 1956. Experiences of 500 children with adult sexual deviation. *Psychiatric Quarterly Supplement* 30: 91–109.

Lane, Harlan. 1975. Constructions of deafness. *Disability and Society* 10: 171–189.

Laqueur, T. W. 1990. *Making Sex: Body and Sex from the Greeks to Freud.* Cambridge, Mass.: Harvard University Press.

Latour, Bruno. 1987. *Science in Action: How to Follow Scientists and Engineers through Society.* Cambridge, Mass.: Harvard University Press.

——— 1988. *The Pasteurization of France.* Cambridge, Mass.: Harvard University Press.

——— 1993. *We Have Never Been Modern.* Cambridge, Mass.: Harvard University Press.

Latour, Bruno, and Steve Woolgar. 1979. *Laboratory Life: The Social Construction of Scientific Facts.* Beverly Hills: Sage.

——— 1986. *Laboratory Life: The Construction of Scientific Facts.* (2d ed. of Latour and Woolgar 1979). Princeton: Princeton University Press. (2d ed. of Latour and Woolgar 1979).

Laudan, Larry. 1977. *Progress and Its Problems: Toward a Theory of Scientific Growth.* Berkeley: University of California Press.

——— 1981. The sources of modern methodology: two models. In L. Laudan, *Science and Hypothesis: Historical Essays on Scientific Methodology.* Dordrecht: Reidel.

——— 1996. *Beyond Positivism and Relativism: Theory, Method and Evidence.* Boulder, Co.: Westview.

Laudan, Rachel. 1987. *From Mineralogy to Geology: The Foundations of a Science 1650–1830.* Chicago: University of Chicago Press.

Leplin, Jarrett. 1984. Introduction. In *Scientific Realism,* ed. J. Leplin. Berkeley and Los Angeles: University of California Press.

Li, C. K., D. J. West, and T. J. Woodhouse. 1990. *Children's Sexual Encounters with Adults.* London: Duckworth.

Lorber, J. 1997. *Gender and the Social Construction of Illness.* Thousand Oaks, Cal.: Sage.

Lorber, J., and Farrell, S. A. 1991. *The Social Construction of Gender.* Newbury Park, Cal: Sage.

Luke, Carmen. 1990. *Constructing the Child Viewer: A History of the American Discourse on Television and Children, 1950–1980.* New York: Praeger.

Lynch, Margaret A. 1985. Child abuse before Kempe. *Child Abuse and Neglect: The International Journal* 9: 7–15.

Lynch, Michael. 1985. *Art and Artifact in Laboratory Science: A Study of Shop Work and Shop Talk in a Research Laboratory.* London: Routledge and Kegan Paul.

——— 1993. *Scientific Practice and Ordinary Action: Ethnomethodology and Social Studies of Science.* Cambridge: Cambridge University Press.

MacCorquodale, K., and Paul D. Meehl. 1948. On a distinction between hypothetical constructs and intervening variables. *Psychological Review* 55: 95–107.

Mack, Pamela Etter. 1990. *Viewing the Earth: The Social Construction of the Landsat Satellite System.* Cambridge, Mass.: MIT Press.

MacKenzie, Donald A. 1981. *Statistics in Britain 1865–1930: The Social Construction of Scientific Knowledge.* Edinburgh: Edinburgh University Press.

——— 1990. *Inventing Accuracy: An Historical Sociology of Nuclear Missile Guidance.* Cambridge, Mass.: MIT Press.

Mannheim, Karl. 1925/1952. *Das Problem einer Soziologie des Wissens.* Trans. in Mannheim, *Essays on the Sociology of Knowledge.* London: Routledge & Kegan Paul.

Margalit, Avishai. 1979. Sense and science. *Essays in Honour of Jaakko Hintikka,* ed. Esa Saarinen et al. Dordrecht: Reidel.

Marshall, Georgina Allen. 1993. *The Social Construction of Child Neglect.* Ph.D. diss. University of British Columbia.

Masson, Jeffrey. 1984. *The Assault on Truth: Freud's Suppression of the Seduction Theory,* New York: Penguin.

May, Larry. 1987. *The Morality of Groups: Collective Responsibility, Group-Based Harm, and Corporate Rights.* Notre Dame: University of Notre Dame Press.

McCormick, Chris. 1995. *Constructing Danger: The Mis/representation of Crime in the News.* Halifax, N. S.: Fernwood.

McHale, Brian. 1992. *Constructing Postmodernism.* New York: Routledge.

McKenzie, Judith A. 1991. The dolomite problem: An outstanding controversy. In *Controversies in Modern Geology,* ed. D. W. Müller, J. A. McKenzie, and H. Weissert. London: Academic Press.

Meadow, R. 1984. Fictitious Illness—the Last Hinterland of Child Abuse. In *Recent Advances in Pediatrics,* ed. R. Meadow. Edinburgh: Oliver and Boyd.

Mehan, Hugh, et al. 1996. *Constructing School Success: The Consequences of Untracking Low Achieving Students.* Cambridge: Cambridge University Press.

Merskey, Harold. 1995. *The Analysis of Hysteria: Understanding Conversion and Dissociation.* 2nd ed. London: Gaskell.

Metalious, Grace. 1956. *Peyton Place.* New York: J. Messner.
Miller, J. B. 1983. *The Construction of Anger in Women and Men.* Wellesley, Mass.: Wellesley College.
Miron, Louis F. 1996. *The Social Construction of Urban Schooling: Situating the Crisis.* Cresskill, N. J.: Hampton Press.
Mitchell, J. V., ed. 1992. *The Eleventh Mental Measurements Yearbook.* Lincoln, Neb.: Buros Institute of Mental Measurements.
Moussa, Helène. 1992. *The Social Construction of Women Refugees: A Journey of Discontinuities and Continuities.* Ed.D. diss., University of Toronto.
Mulkay, Michael Joseph. 1979. *Science and the Sociology of Knowledge.* London: Allen and Unwin.
——— 1991. *Sociology of Science: A Sociological Pilgrimage.* Milton Keynes: Open University Press.
Myers, Greg. 1990. *Writing Biology: Texts in the Social Construction of Scientific Knowledge.* Madison: University of Wisconsin Press.
Nader, Mey, Brenda J. Neff, and Ronald L. Neff. 1986. *Incest as Child Abuse: Research and Applications.* New York: Praeger.
Nanda, Meera. 1997. Against social (de)construction of science: Cautionary tales of the third world. *Monthly Review* 48, no. 10: 1–20.
Nelkin, Dorothy. 1996. What are the science wars really about? *Chronicle of Higher Education* 26, July: A52.
Nelson, Alan. 1994. How *could* scientific facts be socially constructed? *Studies in the History and Philosophy of Science and Technology* 25: 535–547.
Nelson, Barbara. 1984. *Making an Issue of Child Abuse: Political Agenda Setting for Social Problems.* Chicago: Chicago University Press.
Nickles, Thomas. 1988. Reconstructing science: Discovery and experiment. In *Theory and Experiment,* ed. Diderick Batens and Jean Paul van Bendegem. Dordrecht: Reidel.
Nietzsche, Friedrich. 1874/1983. On the uses and disadvantages of history for life. In *Untimely Meditations,* trans. R. J. Hollingdale. Cambridge: Cambridge University Press, 59–123.
Nuttall, Nick, and Nigel Hawkes. 1997. Computer implant gives sight to the blind. *The Times,* 14 September, 8.
Obeyesekere, Gananath. 1992a. *The Apotheosis of Captain Cook: European Mythmaking in the Pacific.* Princeton: Princeton University Press.
——— 1992b. "British cannibals": Contemplation of an event in the death and resurrection of James Cook, explorer. *Critical Inquiry* 18: 630–654.
——— 1997. Afterword: On de-sahlinization. 2nd ed. of Obeyesekere (1992a).
O'Carroll, T. 1980. *Paedophilia: The Radical Case.* London: Peter Owen.
Oliver, Michael. 1990. *The Politics of Disablement.* Basingstoke, Eng.: Macmillan.
O'Neill, Onora. 1989. *Constructions of Reason: Explorations of Kant's Practical Philosophy.* Cambridge: Cambridge University Press.
Parton, Nigel. 1985. *The Politics of Child Abuse.* Basingstoke, Eng.: Macmillan.

Perutz, Max. 1996. Pasteur and the culture wars: An exchange. *New York Review of Books* 46, no. 6; 4 April: 69.

Pfohl, Stephen J. 1977. The "discovery" of child abuse. *Social Problems* 24: 310–323.

Pickering, Andrew. 1984. *Constructing Quarks: A Sociological History of Particle Physics.* Edinburgh: Edinburgh University Press.

——— 1992. *Science as Practice and Culture.* Chicago: University of Chicago Press.

——— 1995. Cyborg history and World War II regime. *Perspectives on Science* 3: 1–48.

——— 1995b. *The Mangle of Practice: Time, Agency, and Science.* Chicago: Chicago University Press.

——— 1995c. Beyond constraint: the temporality of practice and the historicity of knowledge. In *Scientific Practice: Theories and Stories of Doing Physics,* ed. Jed. Z. Buchwald. Chicago: University of Chicago Press.

——— 1997. The mangle as an evolutionary theory of Science and Technology Studies. Physics/History of Science Colloquium: University of Utrecht, Feb. 11, 1997.

Pinch, Trevor. 1986. *Confronting Nature: The Sociology of Solar-Neutrino Detection.* Dordrecht: Reidel.

Pinch, Trevor and Wiebe Bijker. 1987. The social construction of facts and artifacts: or how the sociology of science and the sociology of technology might benefit each other. In Bijker et al. 1987.

Purser, Bruce, et al., eds. 1994. *Dolomites: A Volume in Honour of Dolomieu.* Oxford: Blackwell Scientific.

Putnam, Hilary. 1965. Brains and behaviour. In *Analytical Philosophy,* ed. R. J. Butler. 2nd series, Oxford: Blackwell.

——— 1975. The meaning of "meaning." In Putnam, *Mind, Language and Reality. Vol. 2 of Philosophical Papers.* Cambridge: Cambridge University Press.

——— 1992. *Renewing Philosophy.* Cambridge, Mass.: Harvard University Press.

——— 1994. Sense, nonsense and the senses: An inquiry into the powers of the human mind. *The Journal of Philosophy* 91: 445–517.

Quine, Willard van Orman. 1953. *From a Logical Point of View.* Cambridge, Mass.: Harvard University Press.

——— 1992. *Pursuit of Truth.* Cambridge, Mass.: Harvard University Press.

Radbill, Samuel X. 1968. A history of child abuse and infanticide. In Helfer and Kempe (1968).

Radder, Hans. 1995. In *Scientific Practice: Theories and Stories of Doing Physics,* ed. Jed Z. Buchwald. Chicago: University of Chicago Press.

Raynaud, Maurice. 1862. *Les Médecins au temps de Molière.* Paris: Didier.

Rees, R. van. 1978. Five years of child abuse as a symptom of family problems. In Eekelaar and Katz (1978).

Reynaud, Emmanuel. 1983. *Holy Virility: The Social Construction of Masculinity.* London: Pluto.

Romans, Sarah E., et al. 1993. Otago Women's Health Survey thirty month follow up. I. Onset patterns of non-psychotic psychiatric disorder. II. Remission patterns of non-psychotic psychiatric disorder. *British Journal of Psychiatry* 163: 733–788, 739–746.

Rorty, Richard. 1989. *Contingency, Irony, and Solidarity.* Cambridge: Cambridge University Press.

Rushton, J. Philippe. 1995. *Race, Evolution, and Behavior.* New Brunswick, N.J.: Transaction Publishers.

Russell, Bertrand. 1918. *Mysticism and Logic, and Other Essays.* London: Longman Green.

Russell, Dianne E. H. 1983. The incidence and prevalence of intrafamilial and extrafamilial sexual abuse of female children. *Child Abuse and Neglect: The International Journal* 7: 133–146.

——— 1984. *Sexual Exploitation: Rape, Child Sexual Abuse and Workplace Harassment.* Beverly Hills: Sage.

Sahlins, Marshall. 1981. *Historical Metaphors and Mythical Realities: Structure in the Early History of the Sandwich Islands Kingdom.* Ann Arbor: University of Michigan Press.

——— 1985. *Islands of History.* Chicago: University of Chicago Press.

——— 1995. *How "Natives" Think. About Captain Cook, for Example.* Chicago: University of Chicago Press.

Samson, Colin, and Nigel Smith, eds. 1996. *The Social Construction of Social Policy: Methodologies, Racism, Citizenship and the Environment.* New York: St. Martin's Press.

Savage, Gail. 1996. *The Social Construction of Expertise: The English Civil Service and Its Influence, 1919–1939.* Pittsburgh: University of Pittsburgh Press.

Savelsberg, Joachim J. 1994. *Constructing White-Collar Crime: Rationalities, Communication, Power.* Philadelphia: University of Pennsylvania Press.

Schaffer, Simon. 1993. Letter. *Public Understanding of Science* 2: 264–265.

Scheman, Naomi. 1993. *Engenderings: Constructions of Knowledge, Authority and Privilege.* New York: Routledge.

——— 1997. Types, tokens, and conjuring tricks: Social construction and the reality of the mental. Handout for talk at the University of Toronto, 27 March 1997.

Schultz, L. G. 1982. Child sexual abuse in historical perspective. *Journal of Social Work and Human Sexuality* 1: 21–35.

Schultz, L. G., and P. Jones Jr. 1983. Sexual abuse of children: Issues for social service and health professionals. *Child Welfare* 62: 99–108.

Search, Gay. 1988. *The Last Taboo: Sexual Abuse of Children.* Harmondsworth, Eng.: Penguin.

Searle, J. R. 1995. *The Construction of Social Reality.* New York: The Free Press.

Sgroi, Suzanne. 1975. Sexual molestation of children: The last frontier of child abuse. *Children Today* 4, no. 3:18–21 and continuation.

Shapin, Steven. 1994. *A Social History of Truth.* Chicago: The University of Chicago Press.

——— 1996. *The Scientific Revolution.* Chicago: The University of Chicago Press.

Shapin, Steven, and Simon Schaffer. 1985. *Leviathan and the Air Pump: Hobbes, Boyle and the Experimental Life.* Princeton: Princeton University Press.

Shengold, Leonard. 1989. *Soul Murder: The Effects of Childhood Abuse and Deprivation.* New Haven: Yale University Press.

Shields, Mark A. 1995. *Work and Technology in Higher Education: The Social Construction of Academic Computing.* Hillsdale, N. J.: Lawrence Erlbaum Associates.

Shukla, V., and P. A. Baker, eds. 1988. *Sedimentology and Geochemistry of Dolostones.* Tulsa, Okla.: Society of Economic Paleontologists and Mineralogists Special Publication no. 43.

Sillitoe, Richard H., Robert L. Folk, and Nicolás Saric. 1996. Bacteria as mediators of copper sulfide enrichment during weathering. *Science* 272: 1153–55.

Sismondo, Sergio. 1993. Some social constructions. *Social Studies of Science* 23: 515–553.

——— 1996. *Science without Myth: On Constructions, Reality and Social Knowledge.* Albany: State University of New York Press.

Sokal, Alan. 1996a. Transgressing the boundaries: Toward a transformative hermeneutics of quantum gravity. *Social Text,* Spring/Summer: 217–252.

——— 1996b. A physicist experiments with cultural studies. *Lingua Franca,* May–June: 61–64.

——— 1996c. Response to op-ed of Stanley Fish. *New York Times,* 24 May.

——— 1996d. Transgressing the boundaries: An afterword. *Dissent* 43 (4): 93–99.

Stanley, Liz, and Sue Wise. 1983. *Breaking Out: Feminist Consciousness and Feminist Research.* London: Routledge & Kegan Paul.

Stein, Edward. 1990a. Conclusion: The essentials of constructionism and the construction of essentialism. In Stein (1990b).

Stein, Edward, ed. 1990b. *Forms of Desire: Sexual Orientation and the Social Constructionist Controversy.* New York: Garland. Reprinted in New York: Routledge, 1992.

Tate, Tim. 1990. *Child Pornography: An Investigation.* London: Methuen.

Taylor, Charles. 1971/1985. Interpretation and the sciences of man. In Taylor, *Philosophy and the Human Sciences, Philosophical Papers* 2. Cambridge: Cambridge University Press.

——— 1995. *Philosophical Arguments.* Cambridge, Mass.: Harvard University Press.

Terman, Lewis M., and Maud A. Merrill. 1937. *Measuring Intelligence.* London: Harrap.

Theocharis, T., and Psimopoulos, M. 1987. Where science has gone wrong. *Nature* 329: 595–598.

Tonkin, Elizabeth. 1992. *Narrating Our Pasts: The Social Construction of Oral History.* Cambridge: Cambridge University Press.

Torkington, Ntombenhle Protasia Khotie. 1996. *The Social Construction of Knowledge: A Case for Black Studies.* Liverpool: Liverpool Hope.

Trent, James W. 1994. *Inventing the Feeble Mind: A History of Mental Retardation in the United States.* Berkeley: University of California Press.

Ueno, Chizuko. 1996. Collapse of "Japanese Mothers." *U.S.-Japan Women's Journal,* English Supplement 10:3–19.

Van Fraassen, Bas. 1980. *The Scientific Image.* Oxford: Clarendon Press.

Varga, Donna. 1997. *Constructing the Child: A History of Canadian Day Care Centres.* Toronto: Lorimer.

Vasconcelos, Crisogno, and Judith A. McKenzie. 1997. Microbial mediation of modern dolomitic precipitation and diagenesis under anoxic conditions, Lagoa Vermelha, Rio de Janeiro, Brazil. *Journal of Sedimentary Research* 67: 378–390.

Vasconcelos, Crisogno, Judith A. McKenzie, Stefano Bernasconi, Djorde Grujic, and Alobert J. Tien. 1995. Microbial mediation as a possible mechanism for natural dolomite formation at low temperatures. *Nature* 377; 220–223.

Von Morlot, Adolphe. 1847. Ueber Dolomit und seine künstliche Darstellung aus Kalkstein. *Naturwissenschaftliche Abhandlungen,* ed. W. Hainger. (Vienna: Braunmüller und Seidel) 1: 305–315.

Weber, Ellen. 1977. Incest: Sexual abuse begins at home. *Ms* 5, April: 64–67.

Webster, Christopher D. et al. 1985. *Constructing Dangerousness: Scientific, Legal and Policy Implications.* Toronto: Centre for Criminology, University of Toronto.

Weinberg, Darin. 1997. The social construction of non-human agency: the case of mental disorder. *Social Problems* 44: 217–234.

Weinberg, Steven. 1996a. Sokal's hoax. *New York Review of Books,* August 8: 11–15.

——— 1996b. Reply. *New York Review of Books,* October 3: 55–56.

Weinberg, Thomas S. 1983. *Gay Men, Gay Selves: The Social Construction of Homosexual Identities.* New York: Irvington.

Weisberg, D. Kelly. 1984. The "discovery" of sexual abuse: "Experts" role in legal policy formation. *University of California at Davis Law Review* 18: 1–57.

Wendell, Susan. 1996. *The Rejected Body: Feminist Philosophical Reflections on Disability.* New York: Routledge.

Wertheimer, Alan. 1996. *Exploitation.* Princeton: Princeton University Press.

Wilbur, Cornelia B. 1984. Multiple personality and child abuse: An overview. *Psychiatric Clinics of North America* 7: 3

Wilkins, Robert. 1993. *The Social Construction of the Medicalized Immigrant.* M.A. diss., University of Toronto.

Williams, Bernard. 1978. *Descartes: The Project of Pure Enquiry.* Harmondsworth, Eng.: Penguin.

——— 1981. *Moral Luck.* Cambridge: Cambridge University Press.

——— 1985. *Ethics and the Limits of Philosophy.* Cambridge, Mass.: Harvard University Press.

Wimsatt, William. 1981. Robustness, reliability, and overdetermination. In *Scientific Inquiry and the Social Sciences,* ed. Marilynn B. Brewer and Barry E. Collins. San Francisco: Jossey Bass.

——— 1994. The ontology of complex systems: Levels of organization, perspectives, and causal thickets. In *Biology and Society: Reflections of Methodology. Canadian Journal of Philosophy,* ed. Mohan Matthen and R. X. Ware. Supplementary Vol. 20: 210–274.

Wise, Norton. 1996. Letter. *New York Review of Books,* October 3: 54–55.

Wissow, Lawrence S. 1990. *Child Advocacy for the Clinician: An Approach to Child Abuse and Neglect.* Baltimore: Johns Hopkins.

Wittig, Monique. 1992. *The Straight Mind and Other Essays.* Boston: Beacon Press.

Wolff, Larry. 1988. *Postcards from the End of the World: Child Abuse in Freud's Vienna.* New York: Atheneum.

Wolpert, Lewis. 1993. *The Unnatural Nature of Science.* Cambridge, Mass.: Harvard University Press.

Wong, James. 1997. The "making" of teenage pregnancy. *International Studies in the Philosophy of Science* 11: 273–288.

Woodmansee, Martha, and Peter Jaszi, eds. 1994. *The Construction of Authorship: Textual Appropriation in Law and Literature.* Durham, N. C.: Duke University Press.

Woolgar, Steve, ed. 1988. *Knowledge and Reflexivity.* London: Sage.

Wyatt, G. E., and S. D. Peters. 1986a. Issues in the definition of child sexual abuse in prevalence research. *Child Abuse and Neglect: The International Journal* 10: 231–240.

——— 1986b. Methodological considerations in research on the prevalence of child sexual abuse. *Child Abuse and Neglect: The International Journal* 10: 241–251.

Zenger, Donald H., and S. J. Mazullo, eds. 1982. *Dolomitization.* London: Hutchinson Ross.

Zenger, Donald H., et al. 1994. Dolomieu and the First Description of Dolomite. In Purser et al. (1994).

Ziman, John. 1996. Review of Pickering (1995b). *Metascience.* New Series 9: 40–44.

INDEX